Progress in Systems and Control Theory
Volume 13

Series Editor

Christopher I. Byrnes, Washington University

Associate Editors

S. Balemi
P. Kozák
R. Smedinga
Editors

Discrete Event Systems: Modeling and Control

Proceedings of a Joint Workshop
held in Prague, August 1992

1993

Birkhäuser
Basel · Boston · Berlin

Silvano Balemi
Automatic Control Laboratory
Swiss Federal Institute of
Technology (ETH)
CH-8902 Zürich
Switzerland

Petr Kozák
Inst. of Information Theory
and Automation
Czechoslovak Academy of Sciences
Pod vodárenskou věží 4
182 08 Prague
Czech Republic

Rein Smedinga
Dept. of Computer Science
University of Groningen
P.O. Box 800
NL-9700 AV Groningen
The Netherlands

A CIP catalogue record for this book is available from the Library of Congress,
Washington D.C., USA

Deutsche Bibliothek Cataloging-in-Publication Data

Discrete event systems: modeling and control ; proceedings of
a joint workshop held in Prague, August 1992 / S. Balemi ... ed.
– Basel ; Boston ; Berlin : Birkhäuser, 1993
  (Progress in systems and control theory ; Vol. 13)
    ISBN-13: 978-3-0348-9916-1   e-ISBN-13: 978-3-0348-9120-2
    DOI: 10.1007/978-3-0348-9120-2

NE: Balemi, Silvano [Hrsg.]; GT

© 1993 Birkhäuser Verlag, P.O. Box 133, CH-4010 Basel, Switzerland
Camera-ready copy prepared by the authors in LaTeX
Printed on acid-free paper produced from chlorine-free pulp
Softcover reprint of the hardcover 1st edition 1993

ISBN-13: 978-3-0348-9916-1

9 8 7 6 5 4 3 2 1

CONTENTS

PREFACE

Research of discrete event systems is strongly motivated by applications in flexible manufacturing, in traffic control and in concurrent and real-time software verification and design, just to mention a few important areas. Discrete event system theory is a promising and dynamically developing area of both control theory and computer science.

Discrete event systems are systems with non-numerically-valued states, inputs, and outputs. The approaches to the modelling and control of these systems can be roughly divided into two groups. The first group is concerned with the automatic design of controllers from formal specifications of logical requirements. This research owes much to the pioneering work of P.J. Ramadge and W.M. Wonham at the beginning of the eighties. The second group deals with the analysis and optimization of system throughput, waiting time, and other performance measures for discrete event systems.

The present book contains selected papers presented at the Joint Workshop on Discrete Event Systems (WODES'92) held in Prague, Czechoslovakia, on August 26-28, 1992 and organized by the Institute of Information Theory and Automation of the Czechoslovak Academy of Sciences, Prague, Czechoslovakia, by the Automatic Control Laboratory of the Swiss Federal Institute of Technology (ETH), Zurich, Switzerland, and by the Department of Computing Science of the University of Groningen, Groningen, the Netherlands.

Thirty research papers of high quality were presented and discussed thoroughly by the participants from fourteen countries. Five invited survey lectures given by leading researchers covered main topics of the discrete event system area. Nine different solutions to a simple control problem, stated at the workshop announcement with the aim to stimulate comparison between different approaches, were presented during a special session.

The book reflects four main topics of current research: models of real-time system behaviour, methods for decreasing computational and model complexity, unifying approaches to modelling, and performance analysis of discrete event systems.

The contributions of all speakers and vivid discussions among the participants as well as the quiet summer days in historical Prague contributed to the success of the meeting. We thank the members of the Institute of Information Theory and Automation for their help, in particular Dr. Jiří Pik, Mrs. Marie Kolářová, and Mrs. Jarmila Zoltánová, without whom all this would not have been possible.

We hope that this book containing both recent results and survey papers will be useful as reference material and also as an introductory reading.

The editors

CHAPTER I

LOGICAL MODELS

Augmented Languages and
Supervisory Control
in Discrete Event Systems

Toshimitsu Ushio *

Abstract. We introduce the concept of an augmented language of a specified language. First, we show the relationship between a controllable language and the corresponding augmented language. Next, we give a fixpoint characterization of the augmented language, and propose an algorithm for the computation of a controllable sublanguage for which a finite state supervisor exists. We also deal with blocking in such a finite state supervisor. Finally, we present the relationship between the Wonham-Ramadge algorithm and our proposed algorithm. Moreover, convergence criteria of the Wonham-Ramadge algorithm is extended, and it is shown that there exists a finite state supervisor for the supremal controllable sublanguage under the same condition.

1. INTRODUCTION

Let Σ be a nonempty finite alphabet, and denoted by Σ^* the free monoid over Σ(i.e. the set of all finite strings consisting of elements of Σ, including the empty string ϵ). The concatenation of strings s and t is denoted by st. A string s' is a prefix of a string s if there exists a string s'' with $s = s's''$ [3]. The (prefix) closure $\overline{K}$ of a language K is the language consisting of all the prefixes of strings in K [6]. K is (prefix) closed if $K = \overline{K}$. A complement K^c of K is the set of all strings which are not in K. Languages A and B are said to be nonconflicting if $\overline{A \cap B} = \overline{A} \cap \overline{B}$ [11]. The concatenation AB of languages A and B is the set of all strings s such that $s = s's''$ for some $s' \in A$ and $s'' \in B$. For an automaton H, denoted by $L(H)$ be the language generated by H.

A discrete event system is modeled by an automaton G with an alphabet Σ of events. And let $M = L(G)$. In the supervisory control theory initiated by Ramadge and Wonham [6], Σ is assumed to be decomposed into the disjointed subsets Σ_c and Σ_u of controllable and uncontrollable events respectively. Events in Σ_c are assigned to be enabled or disabled by a supervisor S, and those in Σ_u can not be disabled by it. Denoted by S/G be an automaton representing a closed loop system controlled by S, called a supervised DES. Let L be a specified sublanguage of M or a control specification for G. Note that M is closed and that L is not necessary closed. A language A is said to be $(\Sigma_u, B)-$invariant or simply B-invariant if $\overline{A}\Sigma_u \cap B \subseteq \overline{A}$, where B is a language. Especially, A is said to be controllable if it is M-invariant, and it is shown that there exists a supervisor such that $L(S/G) = \overline{L}$ iff L is controllable [6]. Let $C(L)$ be the set of all controllable sublanguage of L. And $C(L)$ is closed under arbitrary unions,

*Kobe College, Okadayama, Nishinomiya, Hyogo, 662 Japan; e-mail: e00713@sinet.ad.jp

and the supremal element of $C(L)$ under set inclusion, denoted by $\text{sup}C(L)$ or simply $L^\uparrow$, always exists [10].

A supervisor S is usually implemented by a pair of an automaton C and a state feedback Ψ. S is said to be a finite state supervisor if C can be realized by a finite automaton [5]. The finiteness of states of a supervisor is very preferable from the practical point of view because the implementation of a supervisor is very simple. Obviously, a finite state supervisor exists if a specified language is regular, and shown in [7] is an example of a nonregular specified language for which a finite state supervisor exists. A necessary and sufficient condition for the existence of a finite state supervisor has been obtained by Ushio [8]. And it is one of very interesting problems to study an algorithm for the computation of a controllable sublanguage for which there exists a finite state supervisor.

This paper introduces the concept of an augmented language of a specified language, and shows that a finite state supervisor can be designed using an augmented language. Section 2 discusses the relationship between controllability of a specified language and (Σ_u, Σ^*)-invariance of the corresponding augmented language, and shows several properties of (Σ_u, Σ^*)-invariant sublanguage of an augmented language. It is shown that a controllable sublanguage of a specified language for which a finite state supervisor exists is obtained by the supremal (Σ_u, Σ^*)-invariant sublanguage of the corresponding regular augmented language. Also we discuss the possibility of blocking in a finite state supervisor. In section 3, we give a fixpoint characterization of (Σ_u, Σ^*)-invariant augmented languages, and propose an algorithm for the computation of the supremal (Σ_u, Σ^*)-invariant sublanguage of a regular augmented language. We also discuss the relationship between the Wonham-Ramadge algorithm and our proposed algorithm, and extend convergence criteria of the Wonham-Ramadge algorithm. Moreover, it is shown that there exists a finite state supervisor for the supremal controllable sublanguage under the same condition.

2.   AUGMENTED LANGUAGE

Quotient Structure Theorem shown in [6] indicates that a language generated by a supervisor must include the specified language L, and may not disjoint the complement of M. In other words, a superlanguage of L which never includes strings in $M - \overline{L}$ can be considered as a language which does not violate the control specification, that is, an "augmented" specified language. Formally, a superlanguage L_a of L will be called an augmented language of L (with respect to M) if the following conditions hold:

$$L_a \cap M = L \tag{1}$$

$$\overline{L_a} \cap (M - \overline{L}) = \emptyset \tag{2}$$

Let $A(L)$ be a set of all augmented languages of L. Then, as well as $C(L)$, it is easily shown that $A(L)$ is closed under arbitrary unions, and the supremal element of $A(L)$ under set inclusion, denoted by $\text{sup}A(L)$, is given by

$$\text{sup}A(L) = L \cup (\overline{L}\Sigma \cap M^c)\Sigma^*. \tag{3}$$

Note that an augmented language is defined by equation (3) in [9], and the definition is extended in this paper.

First, we show the relationship between controllability and Σ^*-invariance.

Proposition 1 *The following five statements are equivalent.*

(a) *L is controllable.*

(b) *$\sup A(L)$ is Σ^*-invariant.*

(c) *There exists a Σ^*-invariant augmented language of L.*

(d) *There exists an augmented language L_a of L satisfying*

$$\overline{L_a}\Sigma_u \subseteq \overline{\sup A(L)} \tag{4}$$

(e) *Every augmented language L_a of L satisfies equation (4).*

Proof: Three implications (b)$\Rightarrow$(c), (c)$\Rightarrow$(d), and (e)$\Rightarrow$(b) are obvious because $\sup A(L)$ is an augmented language of L. Hence, it is sufficient to prove two implications (d)$\Rightarrow$(a) and (a)$\Rightarrow$(e).

First, we will prove (d)$\Rightarrow$(a). Let L_a be an augmented language satisfying equation (4). Recall that $L_a \cap M = L$. Thus, by equation (3), we have

$$\overline{L}\Sigma_u \cap M \subseteq \overline{L_a}\Sigma_u \cap M \subseteq \overline{\sup A(L)} \cap M = \overline{L},$$

which asserts the statement (a).

Next, we will prove (a)$\Rightarrow$(e). Let L_a be any augmented language of L. Then, we have $\overline{L_a} = \overline{L} \cup L_a'$, where $L_a' \subseteq (\overline{L}\Sigma_u \cap M^c)\Sigma^*$. And

$$\begin{aligned}
\overline{L_a}\Sigma_u &= \overline{L}\Sigma_u \cup L_a'\Sigma_u \\
&= (\overline{L}\Sigma_u \cap M) \cup (\overline{L}\Sigma_u \cap M^c) \cup L_a'\Sigma_u \\
&\subseteq \overline{L} \cup (\overline{L}\Sigma \cap M^c)\Sigma^* \\
&= \overline{\sup A(L)},
\end{aligned}$$

which asserts the statement (e). ∎

A necessary and sufficient condition for the existence of a finite state supervisor is discussed in [8]. The condition can be rewritten in terms of an augmented language as follows.

Proposition 2 *Assume that L is nonempty and controllable. Then there exists a finite state supervisor S such that $L(S/G) = \overline{L}$ if and only if there exists a regular augmented language of $\overline{L}$.*

The following corollary is easily shown from Propositions 1 and 2.

Corollary 1 *Assume that L is nonempty. Then, there exists a finite state supervisor S such that $L(S/G) = \overline{L}$ if and only if there exists a Σ^*-invariant regular augmented language of $\overline{L}$.*

Next, we discuss the relationship between a controllable sublanguage of L and a Σ^*-invariant sublanguage of its augmented language. For a language $N \subseteq \Sigma^*$, let

$$\text{IN}(N) := \{T \, ; \, T \subseteq N \text{ and } \overline{T}\Sigma_u \subseteq \overline{T}.\}.$$

Then, for any $N \subseteq \Sigma^*$, the set $\text{IN}(N)$ is closed under arbitrary unions, and the supremal element of $\text{IN}(N)$ under set inclusion, denoted by $N^{\Uparrow}$, always exists.

Theorem 1 *Let L_a be an augmented language of L, and*

$$L_{c1} := \overline{L_a^{\Uparrow}} \cap M$$

$$L_{c2} := L_a^{\Uparrow} \cap M = L_a^{\Uparrow} \cap L$$

Then the following statements hold.

(1) *L_{c1} is a controllable sublanguage of $\overline{L}$.*

(2) *If $L_a^{\Uparrow}$ and M are nonconflicting, then L_{c2} is a controllable sublanguage of L, and $L_a^{\Uparrow}$ is an augmented language of L_{c2}.*

Proof: Since $L_a^{\Uparrow} \in \text{IN}(L_a)$, the following inclusions hold.

$$\overline{L_a^{\Uparrow}}\Sigma_u \subseteq \overline{L_a^{\Uparrow}} \tag{5}$$
$$L_a^{\Uparrow} \subseteq L_a \tag{6}$$

Equation (6) shows that L_{c1} is a sublanguage of $\overline{L}$. And, by equation (5),

$$\overline{L_{c1}}\Sigma_u \cap M \subseteq \overline{L_a^{\Uparrow}}\Sigma_u \cap M$$
$$\subseteq \overline{L_a^{\Uparrow}} \cap M = \overline{L_{c1}}$$

Therefore, L_{c1} is controllable, which asserts the statement (1).

Next, we will show the statement (2). Obviously, L_{c2} is a sublanguage of L. And

$$\overline{L_{c2}}\Sigma_u \cap M \subseteq \overline{L_a^{\Uparrow}}\Sigma_u \cap M \subseteq \overline{L_a^{\Uparrow}} \cap M. \tag{7}$$

Since M is closed, by the assumption, we have

$$\overline{L_a^{\Uparrow}} \cap M = \overline{L_a^{\Uparrow} \cap M} = \overline{L_{c2}}. \tag{8}$$

Therefore, it is shown by Eqs. (7) and (8) that L_{c2} is controllable. Moreover, by equation (8), we have

$$\overline{L_a^{\Uparrow}} \cap (M - \overline{L_{c2}}) = \emptyset,$$

which implies together with the definition of L_{c2} that $L_a^{\Uparrow}$ is an augmented language of L_{c2}. ∎

Note that L_{c2} is not the supremal controllable sublanguage of L in general even if it is controllable. The following corollary shows that the supremal controllable sublanguage can be computed from a Σ^*-invariant language under certain conditions.

Theorem 2 *Let L_a be an augmented language of L. Assume that L_a is M^c-invariant and $\overline{L}_a \cap M^c \subseteq L_a$. Then the following statements hold.*

(1) *There exists an augmented language L_{ca} of $L^{\uparrow}$ satisfying*

$$L_{ca} \subseteq L_a^{\Uparrow} \tag{9}$$

(2) *If $L_a^{\Uparrow}$ and M are nonconflicting, then*

$$L_a^{\Uparrow} \cap M = L^{\uparrow}, \tag{10}$$

and $L_a^{\Uparrow}$ is an augmented language of $L^{\uparrow}$.

Proof: Let $L_{ca} := L^{\uparrow} \cup (\overline{L^{\uparrow}}\Sigma_u \cap M^c)\Sigma_u^*$. Then L_{ca} is an augmented language of $L^{\uparrow}$, and the following equation holds.

$$\overline{L_{ca}} = \overline{L^{\uparrow}} \cup (\overline{L^{\uparrow}}\Sigma_u \cap M^c)\Sigma_u^* \tag{11}$$

Suppose that $t \in \overline{L^{\uparrow}}$, $\sigma \in \Sigma_u$, and $t\sigma \in M^c$. Then, since L_a is M^c-invariant, we have $\{t\sigma\}\Sigma_u \subseteq \overline{L}_a$. Hence,

$$(\overline{L^{\uparrow}}\Sigma_u \cap M^c)\Sigma_u^* \subseteq \overline{L}_a \cap M^c \subseteq L_a,$$

which implies that $L_{ca} \subseteq L_a$. Moreover, by the controllability of $L^{\uparrow}$ and equation (11), we have

$$\begin{aligned}
\overline{L_{ca}}\Sigma_u &\subseteq (\overline{L^{\uparrow}}\Sigma_u \cap M) \cup (\overline{L^{\uparrow}}\Sigma_u \cap M^c) \cup (\overline{L^{\uparrow}}\Sigma_u \cap M^c)\Sigma_u^* \\
&\subseteq \overline{L^{\uparrow}} \cup (\overline{L^{\uparrow}}\Sigma_u \cap M^c)\Sigma_u^* = \overline{L_{ca}},
\end{aligned}$$

Thus, L_{ca} is Σ^*-invariant, which implies equation (9). We will prove the statement (2). Let $L' := L_a^{\Uparrow} \cap M$. Then, by equation (9), we have $L^{\uparrow} \subseteq L'$. And by Theorem 1, L' is controllable, which asserts the statement (2). ∎

The supremal augmented language of L satisfies the assumptions of the above corollary. Therefore, for any L, there exists an augmented language whose supremal Σ^*-invariant sublanguage is an augmented language of the supremal controllable sublanguage of L.

In general, L_{c2} is not controllable. Obviously, we have $L_{c1} \cap L = L_{c2}$ and $\overline{L_{c2}} \subseteq L_{c1}$, that is, a closed loop system S/G such that $\mathrm{L}(S/G) = L_{c1}$ accepts L_{c2}. Thus, the supervisor may be blocking. Finally, we study a nonblocking condition for such a finite state supervisor. Let S be a supervisor such that $\mathrm{L}(S/G) \subseteq \overline{L}$, and $\mathrm{L}_m(S/G)$ be the set of "marked" strings in S/G, that is,

$$\mathrm{L}_m(S/G) := \mathrm{L}(S/G) \cap L$$

For simplicity, we assume in this paper that $\mathrm{L}_m(G) = \mathrm{L}(G)$. A supervisor S is said to be blocking if $\overline{\mathrm{L}_m(S/G)} \subset \mathrm{L}(S/G)$. This section deals with the issue of blocking in supervisory control in terms of an augmented language. CO

$$\mathrm{L}(S/G) = \overline{L_a^{\Uparrow}} \cap M = L_{c1} \tag{12}$$

and

$$L_m(S/G) = L_a^\Uparrow \cap M = L_{c2} \qquad (13)$$

Theorem 1 indicates that blocking may occur in S/G unless $L_a^\Uparrow$ and M are non-conflicting. We investigate what kinds of blocking can occur in this case. There are two kinds of blocking called the inner and the outer blocking, and two blocking measures $IBM(S)$ and $OBM(S)$ are defined as follows [1]:

$$IBM(S) := (L(S/G) \cap \overline{L}) - \overline{L_m(S/G)}, \qquad (14)$$
$$OBM(S) := L(S/G) - \overline{L}. \qquad (15)$$

Since $\overline{L_a^\Uparrow} \subseteq \overline{L_a}$, we have $\overline{L_a^\Uparrow} \cap M \subseteq \overline{L}$, which implies together with equation (12) that $OBM(S) = \emptyset$, that is, only the inner blocking can occur.

The following lemma is needed to prove inner nonblocking conditions.

Lemma 1 *If the language $L_a^\Uparrow$ is an augmented language of L_{c2}, then $L_a^\Uparrow$ and M are nonconflicting.*

Proof: By the definition of an augmented language, we have

$$\overline{L_a^\Uparrow} \cap (M - \overline{L_a^\Uparrow \cap M}) = \emptyset$$

$$(\overline{L_a^\Uparrow} \cap M) - (\overline{L_a^\Uparrow} \cap \overline{L_a^\Uparrow \cap M}) = \emptyset$$

$$\overline{L_a^\Uparrow} \cap M \subseteq \overline{L_a^\Uparrow} \cap \overline{L_a^\Uparrow \cap M} \subseteq \overline{L_a^\Uparrow \cap M}$$

which asserts that $L_a^\Uparrow$ and M are nonconflicting because $\overline{L_a^\Uparrow \cap M} \subseteq \overline{L_a^\Uparrow} \cap M$. ∎

The following Theorem shows necessary and sufficient conditions for the equality $IBM(S) = \emptyset$.

Theorem 3 *Let S be a supervisor satisfying Eqs. (12) and (13). Then, the following four conditions are equivalent.*

(a) $IBM(S) = \emptyset$.

(b) $L_a^\Uparrow$ is an augmented language of L_{c2}.

(c) $L_a^\Uparrow$ and M are nonconflicting.

(d) $L_a^\Uparrow$ and L are nonconflicting.

Proof: From Lemma 1 and Theorem 1, we can show the equivalence of the conditions (b) and (c). It is sufficient to prove the equivalence of the conditions (c) and (d) and that of (a) and (d).

First, we show the equivalence of (c) and (d). Since $\overline{L} \subseteq M$ and $\overline{L_a^\Uparrow} \cap M \subseteq \overline{L}$, we have

$$\overline{L_a^\Uparrow} \cap \overline{L} \subseteq \overline{L_a^\Uparrow} \cap M \subseteq \overline{L_a^\Uparrow} \cap \overline{L},$$

which implies that

$$\overline{L_a^\Uparrow} \cap \overline{L} = \overline{L_a^\Uparrow} \cap M. \qquad (16)$$

On the other hand, it is obvious that $L_a^\Uparrow \cap L = L_a^\Uparrow \cap M$, and

$$\overline{L_a^\Uparrow \cap L} = \overline{L_a^\Uparrow \cap M}. \tag{17}$$

Therefore, by Eqs. (16) and (17), the conditions (c) and (d) are equivalent.

Next, we show the equivalence of (a) and (d). By the definition of $IBM(\cdot)$,

$$
\begin{aligned}
IBM(S) &= (\mathrm{L}(S/G) \cap \overline{L}) - \overline{\mathrm{L}_m(S/G)} \\
&= (\overline{L_a^\Uparrow} \cap M \cap \overline{L}) - \overline{L_a^\Uparrow \cap L}) \\
&= (\overline{L_a^\Uparrow} \cap \overline{L}) - \overline{L_a^\Uparrow \cap L})
\end{aligned}
$$

Thus the conditions (a) and (d) are equivalent. ■

From Theorems 1 and 3, it is shown that L_{c2} is controllable if S is nonblocking. But the reverse does not hold in general. A counterexample is as follows. Let

$$
\begin{aligned}
\Sigma_u &:= \{\sigma_u\}, \\
\Sigma_c &:= \{\sigma_c\}, \\
M &:= \overline{\sigma_c \sigma_u \sigma_c (\sigma_c + \sigma_u)}, \\
L &:= \overline{\epsilon + \sigma_c \sigma_u \sigma_c^2}, \\
L_a &:= \sigma_u^* + \sigma_c \sigma_u (\sigma_c^2 + \sigma_u^*).
\end{aligned}
$$

Then, $L_a^\Uparrow = \sigma_u^* + \sigma_c \sigma_u \sigma_u^*$, and $L_{c2} = \{\epsilon\}$. Obviously, L_{c2} is controllable but $L_a^\Uparrow$ is not an augmented language of L_{c2}. Thus, using $L_a^\Uparrow$, we can design a finite state supervisor S satisfying Eqs. (12) and (13) when L_a is regular. But blocking occurs in S/G.

3. COMPUTATION OF $L_a^\Uparrow$

This section shows an algorithm for the computation of the supremal Σ^*-invariant sublanguage of an augmented language L_a of L. First we define an operator Ω from the class of languages over Σ into itself as follows:

$$
\begin{aligned}
\Omega(N) &:= L_a \cap \sup\{\, T\,;\, T \subseteq \Sigma^*,\ T = \overline{T} \text{ and } T\Sigma_u \subseteq \overline{N} \,\} \\
&= \{\, t\,;\, t \in L_a \text{ and } \overline{t}\Sigma_u \subseteq \overline{N} \,\}, \tag{18}
\end{aligned}
$$

where $\overline{t} = \overline{\{t\}}$. A language N is said to be a fixpoint of Ω if $\Omega(N) = N$.

Proposition 3 *Every Σ^*-invariant sublanguage of L_a is a fixpoint of Ω. And for any fixpoint T of Ω, the following inclusion holds.*

$$T \subseteq L_a^\Uparrow. \tag{19}$$

The proof of the above proposition is similar to that of Proposition 2. 1 in [10]. Next, we consider a sequence $\{\, K_j\,\}_{j=0}^\infty$ of languages defined by

$$
\begin{aligned}
K_0 &= L_a \\
K_{j+1} &= \Omega(K_j) \qquad j=0,1,\dots \tag{20}
\end{aligned}
$$

Obviously, Ω is monotone, that is, $\Omega(T) \subseteq \Omega(S)$ whenever $T \subseteq S \subseteq \Sigma^*$. Thus The following proposition is easily shown from Proposition 3 and equation (20).

Proposition 4 *For the sequence* $\{ K_j \}_{j=0}^{\infty}$ *of languages defined by equation (20), the limit* K_{∞} *exists. And*

$$L_a^{\Uparrow} \subseteq K_{\infty} \tag{21}$$

In general, the reverse inclusion of equation (21), that is, $L_a^{\Uparrow} = K_{\infty}$ does not hold. Then, We restrict the case that L_a is regular. For a language N, strings s and $t \in \Sigma^*$ are said to be equivalent (mod N), denoted by $s \equiv_N t$ if

$$\{ s' \, ; \, s' \in \Sigma^* \text{ and } ss' \in N \} = \{ t' \, ; \, t' \in \Sigma^* \text{ and } tt' \in N \}.$$

And let $\| N \|$ be the cardinal number of equivalence classes of $\equiv_N$ in Σ^*. Note that N is regular if and only if $\| N \|$ is finite [3]. Then the following theorem can be easily shown.

Theorem 4 *Assume that* L_a *is regular. Then the sequence* $\{ K_j \}_{j=0}^{\infty}$ *of languages defined by equation (20) converges to* $L_a^{\Uparrow}$ *after a finite number of iterations. And* $L_a^{\Uparrow}$ *is a regular language satisfying*

$$\| L_a^{\Uparrow} \| \leq \| L_a \| + 1 \tag{22}$$

The proof of the above theorem is similar to that of Theorem 3.1 in [10], and omitted. By Proposition 2, Theorems 1, 2, and 4, we show the following corollary.

Corollary 2 *Assume that there exists a regular augmented language* L_a *of* L. *Let* $L_c := L_a^{\Uparrow} \cap M$. *If* $L_a^{\Uparrow}$ *and* M *are nonconflicting, then* L_c *is a controllable sublanguage of* L, *and there exists a finite state supervisor* S *such that* $L(S/G) = \overline{L_c}$. *Moreover, if* L_a *is* M^c*-invariant and* $\overline{L_a} \cap M^c \subseteq L_a$, *then* L_c *is the supremal controllable sublanguage of* L.

By Wonham and Ramadge [10], an algorithm for the computation of the supremal controllable sublanguage has been proposed, and its convergence after a finite number of iterations has been shown in the case that both L and M are regular. Note that there may exist a regular augmented language when either L or M are nonregular. And in such a case, the supremal controllable sublanguage can be obtained by a finite number of iterations of equation (20) under assumptions of Corollary 2.

Finally, we discuss the relationship between the Wonham-Ramadge algorithm and our proposed algorithm. The Wonham-Ramadge algorithm is defined as follows [10]:

$$\begin{aligned} K_0' &= L \\ K_{j+1}' &= \Omega'(K_j'), \quad \text{j=1,2,} \ldots \end{aligned} \tag{23}$$

where

$$\begin{aligned} \Omega'(K) &= L \cap \sup\{T \, ; \, T = \overline{T}, \, T \subseteq \Sigma^* \text{ and } T\Sigma_u \cap M \subseteq \overline{K}\} \\ &= \{t \, ; \, t \in L \text{ and } \overline{t}\Sigma_u \cap M \subseteq \overline{K}\}. \end{aligned}$$

Lemma 2 *If* L_a *is* M^c*-invariant and closed, then* K_j *for* $j = 0, 1 \ldots$ *is* M^c*-invariant and closed.*

Proof: Since $K_0 = L_a$, K_0 is M^c-invariant and closed. Thus, it is obvious that K_j is closed for each $j = 0, 1 \ldots$. For the inductive step, suppose that K_i is M^c-invariant. Then, for each $t \in K_{i+1}\Sigma_u \cap M^c$, there exist $t' \in K_{i+1}$ and $\sigma_u \in \Sigma_u$ such that $t = t'\sigma_u$, and we have $\overline{t'}\Sigma_u \subseteq K_i$. Thus, it is sufficient to show that $\{t\}\Sigma_u \subseteq K_i$. Since $\{t\}\Sigma_u \in M^c$ and $t \in K_i$, we have

$$\{t\}\Sigma_u = \{t\}\Sigma_u \cap M^c \subseteq K_i\Sigma_u \cap M^c \subseteq K_i$$

Therefore, $t \in K_{i+1}$, and K_{i+1} is M^c-invariant. ∎

The following Theorem shows the relationship between the Wonham-Ramadge algorithm (23) and our algorithm (20).

Theorem 5 *Let $\{K_j\}$ and $\{K_j'\}$ be the sequences defined by Eqs. (20) and (23) respectively. Assume that L is closed, and L_a is M^c-invariant and closed. Then,*

$$K_j \cap M = K_j' \qquad for \ j = 0, 1, \ldots \qquad (24)$$

Proof: For $j = 0$, we have $K_0 \cap M = L_a \cap M = L = K_0'$. For the inductive step, suppose that $K_i \cap M = K_i'$. Let $t \in K_{i+1} \cap M$. Then, $t \in L_a \cap M = L$, and by lemma 2,

$$\overline{t}\Sigma_u \subseteq \overline{K_i} = K_i.$$

And

$$\overline{t}\Sigma_u \cap M \subseteq K_i \cap M = K_i.$$

Thus, we have $t \in K_{i+1}'$, and $K_{i+1} \cap M \subseteq K_{i+1}'$.
On the other hand, $K_{i+1}' \subseteq M$, and

$$\overline{K_{i+1}'\Sigma_u} \cap M \subseteq \overline{K_i'}.$$

And, by Lemma 2,

$$\begin{aligned}
\overline{K_{i+1}'\Sigma_u} &= (\overline{K_{i+1}'\Sigma_u} \cap M) \cup (\overline{K_{i+1}'\Sigma_u} \cap M^c) \\
&\subseteq \overline{K_i'} \cup (\overline{K_i'\Sigma_u} \cap M^c) \\
&= (K_i \cap M) \cup (\overline{K_i'\Sigma_u} \cap M^c) \\
&\subseteq K_i \cup (\overline{K_i'\Sigma_u} \cap M^c) \\
&\subseteq K_i \cup (\overline{K_i\Sigma_u} \cap M^c) \\
&\subseteq K_i = \overline{K_i}.
\end{aligned}$$

Therefore, $K_{i+1}' \subseteq K_{i+1}$, which implies together with $K_{i+1}' \subseteq M$ that $K_{i+1}' \subseteq K_{i+1} \cap M$. ∎

The following theorem is easily shown by Corollary 1, Theorems 1, 4, and 5.

Theorem 6 *Suppose that L is closed and that there exists an M^c-invariant closed regular augmented language of L. Then equation (23) converges after a finite number of iteration to $\sup C(L)$. And there exists a finite state supervisor for $\sup C(L)$.*

Note that there exists an augmented language satisfying the assumption of Theorem 6 if both L and M are regular, but that the reverse is false. Hence, Theorem 6 is a generalization of Theorem 3.1 in [10].

4. CONCLUSIONS

We have introduced the concept of an augmented language of a specified language, and studied its application to the design of a finite state supervisor. Using an augmented language, we can compute a controllable (possibly nonregular) sublanguage for which a finite state supervisor exists even if M and L are nonregular. However, blocking may occur in the closed loop system controlled by such a supervisor. If L_{c2} is controllable, we can design a nonblocking supervisor S' such that $L(S'/G) = \overline{L_{c2}}$, but, in general, S' may not be a finite state one, that is, nonblocking and finiteness of states of a supervisor may be alternative. It is a future work to obtain necessary and sufficient conditions for S' to be a finite state supervisor.

Unfortunately, There exists no algorithm for the computation of a regular augmented language in general since M and L are not regular. It is one of a very interesting problem to study heuristic methods for finding an augmented language by restricting classes of a controlled discrete event system and a specified language.

REFERENCES

[1] E. Chen and S. Lafortune, "Dealing with blocking in supervisory control of discrete-event systems," *IEEE Trans. on Automatic Control,* vol. AC-36, no. 6, pp. 724-735, 1991.

[2] E. Chen and S. Lafortune,, "On nonconflicting languages that arise in supervisory control of discrete event systems," *Systems & Control Letters,* vol. 17, , pp. 105-113, 1991.

[3] J. E. Hopcroft and J. D. Ullman, *Introduction to Automata Theory, Language, and Computation,* Addison-Wesley, Reading MA, 1979.

[4] S. Lafortune and E. Chen, "The infimal closed controllable superlanguage and its application in supervisory control," *IEEE Trans. on Automatic Control,* vol. AC-35, no. 4, pp. 398-405, 1990.

[5] P. J. Ramadge, "Some tractable supervisory control problems for discrete-event systems modeled by Büchi automata," *IEEE Trans. Automatic Contr.,* vol. AC-34, no. 1, pp. 10-19, 1989.

[6] P. J. Ramadge and W. M. Wonham, "Supervisory control of a class of discrete-event processes," *SIAM J. Contr. Optimiz.,* vol. 25, pp. 206-230, 1987.

[7] T. Ushio, "On controllability of controlled Petri nets," *Control - Theory and Advanced Technology,* vol. 5, no. 3, pp. 265-275, 1989.

[8] T. Ushio, "On the existence of finite state supervisors in discrete-event systems," in *Proc. 29th IEEE CDC.,* pp. 2857-2860, 1990.

[9] T. Ushio, "The augmented language and its application in supervisory control," *Recent Advances in Mathematical Theory of Systems, Control, Networks and Signal Processing II,* edited by H. Kimura and S. Kodama, pp. 43-48, 1992.

[10] W. M. Wonham and P. J. Ramadge, "On the supremal controllable sublanguage of a given language," *SIAM J. Contr. Optimiz.,* vol. 25, pp. 637-659, 1987.

[11] W. M. Wonham and P. J. Ramadge, "Modular supervisory control of discrete-event systems," *Math. Contr. Signals and Syst.,* vol. 1, no. 1, pp. 13-30, 1988.

Input/Output Discrete Event
Processes and System Modeling

Silvano Balemi*

Abstract. In this paper we present an input/output interpretation of supervisory control theory. As opposed to the model proposed by Ramadge and Wonham, the plant does not spontanuously generate all events as outputs, but reads some events, called commands, as inputs and produces other events, called responses, as outputs. Based on this input/output interpretation we propose a modeling framework that starts with a physical description of the system and ends with the system model. This is performed by designing software routines interacting with the system at appropriate time instances.

1. INTRODUCTION

The supervisory control theory introduced by Ramadge and Wonham [1] provides a framework for the treatment of theoretical and computational issues related to the control of discrete event systems. However, the interpretation and implementation of the results obtained in this research area are not always straightforward. This is partly due to the model semantics of a plant generating events with a supervisor only passively preventing the generation of some of the events.

This semantics is not appropriate for modeling many real life systems because it does not naturally allow the supervisor to generate events of its own. A partial step in the direction of a supervisor able to force the occurrence of certain events in the plant was done by Golaszewski and Ramadge [2]. In this paper we use an explicit input/output semantics for the plant. The plant reads inputs, called commands, and produces outputs, called responses. Symmetrically, a controller accepts as inputs the responses produced by the plant and generates as outputs the commands for the plant.

A crucial step toward the practical control of a discrete event system is the modeling step. As opposed to the modeling of continuous-valued systems however, the task of modeling a discrete event system is not a "passive" process in which one selects the best suitable model from a class of models according to some given criteria. A model for a discrete event system must not be "identified" but designed and constructed.

This is performed by adding to the given system some additional structure, for instance in form of some software that interacts with the (physical) system via actuators and sensors. Following the I/O semantics above, the inputs correspond to the call of software routines, while the outputs are messages reporting the qualitative changes occurred in the system.

*Automatic Control Laboratory, Swiss Federal Institute of Technology (ETH), Zurich, Switzerland; e-mail: balemi@aut.ee.ethz.ch

In this paper we first introduce the input/output control problem [3, 4, 5]. Then we present a modeling procedure for obtaining an input/output model of the considered discrete event system. We illustrate this with the example of a valve of a semiconductor manufacturing piece of equipment [4, 5].

1.1 Preliminaries and Notation

The behavior of discrete event systems is naturally described by sequences (called *trace*) of certain qualitative changes (called *events*) in the system, by ignoring micro-changes occurring between two events. Ramadge and Wonham introduced a language-theoretical approach on control of DES by considering the possible traces to be strings on an alphabet of symbols representing the events. The set of all such strings is then called a *language*, and represents the possible behavior of the system. We denote the alphabet of symbols by Σ. The quantity Σ^* stands for the set of all finite sequences (or strings) over Σ. The empty string is denoted by ϵ. For a language $L \subset \Sigma^*$, $\overline{L}$ denotes the set of prefixes of strings in L. We say a language L is *prefix-closed* if $L = \overline{L}$.

1.2 Process Model and Process Composition

A process is a triple $P = (\Sigma, L_P, M_P)$ composed of two languages L_P and M_P of finite traces over the alphabet Σ. L_P is prefix-closed and represents all partial traces of P, while $M_P \subseteq L_P$, is a set of distinguished traces, the *marked language*. Often M_P marks the set of successfully *completed* traces.

An important operator on processes is the *prioritized synchronous composition* [6]. Before defining it, we need additional notation. The *generalized projection* of a string s onto a prefix-closed language L, denoted by $\mathrm{gproj}[L](s)$, is a string defined recursively by $\mathrm{gproj}[L](\epsilon) = \epsilon$ and with $\sigma \in \Sigma$ by

$$\mathrm{gproj}[L](s.\sigma) = \begin{cases} \mathrm{gproj}[L](s).\sigma & \text{if } \mathrm{gproj}[L](s).\sigma \in L \\ \mathrm{gproj}[L](s) & \text{otherwise.} \end{cases}$$

The generalized projection of a string onto a language yields the string that is obtained from the original string by discarding in a row those letters that could not "follow" language L. The expression $\mathrm{gproj}[L](K)$ denotes the obvious extension to a language K. For an arbitrary alphabet Σ', the special case $\mathrm{gproj}[(\Sigma')^*](K)$ is called the *natural projection of K on the alphabet Σ'*. We are now ready to define the composition operator.

Definition 1 (Prioritized Synchronous Composition) *The prioritized synchronous composition $P_{1\,A}\|_B P_2$ of $P_1 = (\Sigma_1, L_{P_1}, M_{P_1})$ and $P_2 = (\Sigma_2, L_{P_2}, M_{P_2})$ with respect to the sets $A \subseteq \Sigma_1$ and $B \subseteq \Sigma_2$ is*

$$P_{1\,A}\|_B P_2 = (\Sigma_1 \cup \Sigma_2, \ L_{P_{1\,A}\|_B P_2}, \ M_{P_{1\,A}\|_B P_2}) \tag{1}$$

where $L_{P_{1\,A}\|_B P_2}$ is defined by $\epsilon \in L_{P_{1\,A}\|_B P_2}$ and by the recursive condition that for $s \in L_{P_{1\,A}\|_B P_2}$ and $\sigma \in \Sigma_1 \cup \Sigma_2$ the string $s.\sigma$ is in $L_{P_{1\,A}\|_B P_2}$ only if the following

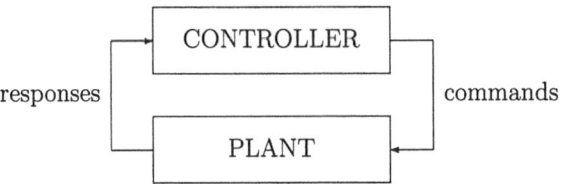

Figure 1: Closed-loop system

condition is satisfied:

$$\begin{array}{ll}
\mathrm{gproj}[L_{P_1}](s).\sigma \in L_{P_1} & \forall \sigma \in A - B \\
\mathrm{gproj}[L_{P_2}](s).\sigma \in L_{P_2} & \forall \sigma \in B - A \\
\mathrm{gproj}[L_{P_1}](s).\sigma \in L_{P_1} \ \wedge \ \mathrm{gproj}[L_{P_2}](s).\sigma \in L_{P_2} & \forall \sigma \in A \cap B \\
\mathrm{gproj}[L_{P_1}](s).\sigma \in L_{P_1} \ \vee \ \mathrm{gproj}[L_{P_2}](s).\sigma \in L_{P_2} & \forall \sigma \notin B \cup A.
\end{array} \qquad (2)$$

The marked language $M_{P_1 {}_A\|_B P_2}$ is described by

$$M_{P_1 {}_A\|_B P_2} = \left\{ s \in L_{P_1 {}_A\|_B P_2} \mid \mathrm{gproj}[L_{P_1}](s) \in M_{P_1} \ \wedge \ \mathrm{gproj}[L_{P_2}](s) \in M_{P_2}. \right\}$$

The sets A and B are called the priority sets for process P_1 and P_2 respectively. They indicate the events that can occur in the composition only if the respective process agrees on their execution. The events not in the priority set of a process can be executed by the other process at any time. The special case of $A = \Sigma_1$ and $B = \Sigma_2$ is called the *full synchronous composition*, and is denoted by $P_1 \| P_2$; the processes must execute simultaneously a common event (in $\Sigma_1 \cap \Sigma_2$), while the other events can happen independently in each respective process. The other special case of $A = B = \emptyset$ is called *parallel composition*; the processes must execute a common event if both can, otherwise they are free to execute any event independently. The parallel composition of two processes P_1 and P_2 is denoted by $P_1 /\!/ P_2$.

2.   AN INPUT/OUTPUT SEMANTICS

We introduce here an input/output model of discrete event systems. The alphabet of the process languages is divided into two disjoint subsets: the commands (denoted by Σ_c) and the responses (denoted by Σ_r). As proposed in [3], the set Σ_c of commands models the inputs of the plant process whereas the set Σ_r of responses stands for the plant outputs.

The basic problem we want to address is the enforcement of a behavior on the plant $P = (\Sigma, \ L_P, \ M_P)$. We do this with a process $C = (\Sigma, \ L_C, \ M_C)$ called controller. Symmetrically to the plant, the controller accepts responses as inputs and produces commands as outputs. Then, the closed-loop behavior of the plant and the controller connected together as in figure 1 can be described by the full synchronous composition of the two processes. The resultings process is then $C \| P = C_\Sigma \|_\Sigma P = (\Sigma, \ L_{P\,\Sigma} \|_\Sigma L_C, \ M_{P\,\Sigma} \|_\Sigma M_C)$.

For the closed-loop, we require that any response sent by the plant can be accepted by the controller. This is expressed by the condition

$$P\|C = P_\Sigma\|_{\Sigma_c}C. \tag{3}$$

Controllers satisfying this constraint can never inhibit by synchronization the occurrence of a response in the plant. This condition can be rephrased using the notion of controllability of a language [7]. A language $K \subset \Sigma^*$ is called *controllable with respect to* L *and* Σ' if the inclusion $\overline{K}.\Sigma' \cap \overline{L} \subseteq \overline{K}$ holds. Then it can be shown [8] that a controller satisfies equation (3) if and only if L_C is controllable w.r.t. L_P and Σ_r *i.e.* if and only if $L_C.\Sigma_r \cap L_P \subseteq L_C$ holds. Symmetrically to the previous condition, we also require in the closed-loop that any command sent by the controller can be accepted by the plant. Thus the controller does not need to check if the plant can accept the command but knows a priori that the plant can. This is expressed by the following condition

$$P\|C = P_{\Sigma_r}\|_\Sigma C. \tag{4}$$

As for equation (3), condition (4) holds if and only if L_P is controllable with respect to L_C and Σ_c, *i.e.* if and only if $L_P.\Sigma_c \cap L_C \subseteq L_P$ holds. The closed-loop connection of a controller and a plant satisfying the two conditions above is said to be *well-posed*. An output produced by any of the processes in the closed loop can be accepted as an input by the other process. This is in analogy to the concept of well-posedness in control theory, where it indicates that the connection of some systems "makes sense".

In addition to the condition of well-posedness, we are interested in controllers that always guarantee the termination of a marked string. For this, Ramadge and Wonham [9] introduced the concept of non-blockingness. A process P is non-blocking if $L_P = \overline{M_P}$. We are now ready to present the input/output control problem.

Input/Output Control Problem

Given a plant P *and a specification language* $L_{\text{spec}}^{\min}$, $L_{\text{spec}}^{\max} \subseteq \Sigma^*$ *for the closed-loop behavior, find a controller* C *such that*

1. $L_{\text{spec}}^{\min} \subseteq M_P^c \subseteq L_{\text{spec}}^{\max}$,
2. $C\|P$ *is non-blocking*,
3. *The connection of* P *and* C *is well-posed.*

The solution of the control problem can be expressed using the notion of controllability. The class $\mathcal{C}(L)$ of sublanguages of a language L which are controllable with respect to L_P and Σ_r is closed under language union, and therefore there is a *supremal* element of $\mathcal{C}(L)$, denoted by $\sup \mathcal{C}(L)$. It can be shown [4, 5] that the input/output control problem has a solution if and only if $L_{\text{spec}}^{\min} \subseteq \sup \mathcal{C}(M_P \cap L_{\text{spec}}^{\max})$. If a solution exists, the controller

$$C_{\text{sup}} = (\Sigma, \overline{\sup \mathcal{C}(M_P \cap L_{\text{spec}}^{\max})}, \sup \mathcal{C}(M_P \cap L_{\text{spec}}^{\max})) \tag{5}$$

is always a solution.

3. MODELING PRINCIPLES

In this section we present a methodology for obtaining an input/output model. The construction of a model is illustrated for a gas valve component of a semiconductor manufacturing piece of equipment: the rapid thermal multiprocessor (RTM) at the Center for Integrated Systems of Stanford University [4, 5].

The task of modeling a discrete event system can be decomposed into four parts. While not all systems are easily modeled by using this methodology, the steps described below have proven to be helpful for modeling the RTM and are general enough to be of widespread practicality.

Step 1 – Model the high-level behavior of the plant. This results in the description of a fundamental process Ξ that best describes the possible functions of the (physical) plant.

Step 2 – Design a logical interface to the physical plant, starting from the fundamental process Ξ. This consists of designing low-level routines that interact with the physical system by triggering actuators or reporting sensor readings. The routines are represented by *actuating* and *sensing* processes Ψ_i.

Step 3 – Compose the description of the fundamental process with the above routines. This results in a candidate process model P' for the plant.

Step 4 – Choose a subprocess P of the process P' found in step 3 so that some desired properties of the plant are satisfied.

The fundamental process $\Xi = (\Sigma', L_\Xi, M_\Xi)$ models the qualitative changes that can occur in the system under consideration. It can be thought of as the fundamental behavior that is consistent with the physical equipment to be controlled. It represents the abstract view of the functionalities of the system.

At a more detailed level of modeling, each event in the fundamental process corresponds to a sequence of events taking place in the *actuating* and *sensing* processes. These underlying sequences of events are modeled by p separate processes or routines, $\Psi_i = (\Sigma_i, L_{\Psi_i}, M_{\Psi_i})$, $1 \le i \le p$, with the languages L_{Ψ_i} consisting of a command followed by a finite number of responses, *i.e.* $L_{\Psi_i} \subseteq \Sigma_c.\Sigma_r^*$ (Σ_c and Σ_r are the sets of all commands resp. responses in the sets Σ_i). The initial command can be thought of as the call of the routine, while the responses as its effects. Note that the languages M_{Ψ_i} are not necessarily prefix-closed, *i.e.* $M_{\Psi_i} \subset L_{\Psi_i}$. In particular, we suppose that the only markings are the strings in L_{Ψ_i} without possible continuation, *i.e.* $M_{\Psi_i} = \{s \in L_{\Psi_i} \mid s.\Sigma \cap L_{\Psi_i} = \emptyset\}$.

The interaction of fundamental process and routines can be explained in the following way. A command corresponding to an available routine is given and, after the command, the fundamental process starts producing responses from the routines according to its own possible behavior described by the language L_Ξ. The routines that have been started with a command and that have not yet ended, *i.e.* that have not yet reached a marking, are called *running* and they can be activated again only after having reached a marking.

A candidate process model for the input/output plant is the process $P' = (\Sigma,\ L_{P'},\ M_{P'})$ given by

$$P' = \Xi \parallel (\mathbin{/\!\!/}_{i=1}^{p} \Psi_i^*) \tag{6}$$

with the alphabet $\Sigma = \Sigma' \cup \Sigma_1 \cup \Sigma_2 \cdots \cup \Sigma_p$.

The process Ψ_i^* (see also the sequential composition of Inan and Varaiya [10]) is defined by

$$\Psi_i^* = (\Sigma_i,\ \overline{M_{\Psi_i}^*},\ M_{\Psi_i}^*)$$

where the symbol * for the languages in this equation stands for the extension of the Kleene closure to languages. Given a marked language M, this means that $s \in M^*$ if and only if there exist strings $s_1,\ s_2 \cdots s_m \in M$, for some m such that $s_1.s_2.\cdots.s_m = s$.

In equation (6), the processes Ψ_i^* model the repetitive execution of the routines Ψ_i. Note, however, that this repetitive execution must occur for each individual routine in a sequential fashion only. The routines ask the use of some resource, and must possibly wait until the fundamental process is ready to make the resource available. This is therefore expressed by the *synchronous composition* of the fundamental process Ξ with the process $\mathbin{/\!\!/}_{i=1}^{p} \Psi_i^*$ resulting from the parallel composition of the repeated routines Ψ_i^*.

3.1   Nonblocking Plants with Terminating Routines

On the plant process obtained in equation (6) we want to enforce an additional constraint. We require that all the running routines can always spontaneously terminate by reaching a marking, *i.e.* the following condition must be satisfied for a process $P = (\Sigma,\ L_P,\ M_P)$

$$\forall s \in L_P\ \exists t \in \Sigma_r^*\ |\ \mathrm{gproj}[\overline{M_{\Psi_i}^*}](s.t) \in M_{\Psi_i}^*\quad \forall\ i = 1,..,p. \tag{7}$$

The class of languages satisfying equation (7) is closed under union as stated in the next proposition.

Proposition 1 *The set of languages satisfying equation (7) is closed under union of languages. Then, given L_P' as the language of the process P' obtained from equation (6), there exists a supremal sublanguage of L_P' satisfying equation (7). Moreover, this supremal sublanguage is controllable w.r.t. L_P' and Σ_r.*

Proof: The proof is easy by noting that the class of languages satisfying equation (7) is closed under union because the property can be checked for each string in $M_{\Psi_i}^*$ or prefix individually. From this property, the existence of a supremal sublanguage of L_P satisfying equation (7) is automatically proved. The controllability w.r.t. L_P' and Σ_r of this language follows from the definition of the languages M_{Ψ_i} which have an element of Σ_c as first element of any string, and then a sequence of elements from Σ_r.   ∎

This proposition implies that there exists a least restrictive plant P satisfying equation (7), *i.e.* a plant allowing to start as many routines as possible under the constraint given by equation (7).

Then, we can use P as a model for the plant (combining fundamental process and routines) because by choosing the model we only limit a priori the availability of the routines that may be or may not be started in the candidate plant P'. While every routine that is not running can be usually started at any time with a command in the process P', the enforcement of equation (7) requires that a routine be initiated only if this routine and all the routines currently running can eventually terminate by reaching a marking.

The language of the least restrictive plant can be determined as follows. Consider any command in the candidate plant P'. If after a string s in $L_{P'}$ and this command it is possible to reach a point where a routine cannot spontaneously terminate, then the continuation of the string with the command should be removed from the language.

This can be formulated as follows. Take $L_P = L_{P'}$ and consider any string $s \in L_P$ and $\sigma_c \in \Sigma_c$, then

$$\text{if } \exists\, i \mid \overline{\text{gproj}[\overline{M_{\Psi_i}^*}](s.\sigma_c.\Sigma_r^*)} \cap \overline{M_{\Psi_i}^*} \neq \overline{\text{gproj}[\overline{M_{\Psi_i}^*}](s.\sigma_c.\Sigma_r^*)} \cap M_{\Psi_i}^*$$

$$\Rightarrow \text{ remove } \{s.\sigma_c.\Sigma^*\} \text{ from } L_P$$

This should be repeated for every command σ_c in any string s originally contained in $L_{P'}$. The resulting L_P is the language of a plant $P = (\Sigma,\ L_P,\ L_P \cap M_{P'})$ satisfying equation (7).

Note that this plant is not necessarily non-blocking because all running routines could have terminated while the fundamental process is not marked. Then the non-blocking plant P_{nb} can be determined using the notion of controllability. In fact the least restrictive non-blocking plant satisfying equation (7) is given by

$$P_{nb} = (\Sigma,\ \overline{M_{P_{nb}}},\ M_{P_{nb}})$$

where $M_{P_{nb}}$ is the supremal sublanguage of $L_P \cap M_{P'}$ which is controllable w.r.t. to L_P and Σ_r. Note also that with the step from P to P_{nb} the property given by equation (7) is preserved.

3.2 Plants subject to Communication delays

In real-life systems the message exchanges between the plant and the controller are affected by communication delays. A message sent by a process needs some time to reach the other process. Besides, the time needed for the commands to be executed can be modeled as a delay in the transmission of the responses.

Because of the communication delays, there can be inconsistencies between the controller and the plant. For instance, the controller could send a command to the plant without the knowledge of a response already produced by the plant but not yet arrived at the controller input.

Then, in order to enforce a desired behavior on a plant in a closed-loop affected by communication delays, more sophisticated supervisors/controllers must be used in general. Li and Wonham [11] (in the original setup of Ramadge and Wonham) and the author [12] (in an I/O-setup) discuss the properties of such supervisors/controllers.

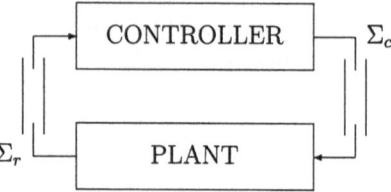

Figure 2: Closed-loop connection of plant and controller processes with communication delays.

More sophisticated controllers, however, are not necessary if the plant satisfies additional properties. For instance, if the plant is such that no command can be given while some routine is running, *i.e.* if no element of Σ_c can be accepted by the plant after a strings s such that $\text{gproj}[\overline{M^*_{\Psi_i}}](s) \notin M^*_{\Psi_i}$ for some routine index i, then a controller solving the problem on page 18, correctly enforces the specification on the plant affected by communication delays. Such plants trivially behave correctly under delays because the controller is forced to wait until *the current* routine has terminated.

3.3 Two-stage Design

Therefore we can say that a model for the plant (the control model) is *designed* from the physical description of the system. The aim of this design is to obtain a plant model ensuring desired operating constraints.

Given the plant model, the second step consists in designing the controller which enforces given specifications on the plant's closed-loop behavior. We can see these two design steps in figure 3.

As we showed for the case of a plant-controller pair affected by communication delays, the design of a controller can be performed in a less computationally intensive way and using known algorithms if the plant satisfies given properties.

Plant design versus controller design. Now we could ask ourselves if it makes sense to invest more time in the design of the model instead of determining a general model and then take care with the controller design of both operational constraints and control specifications. A partial list of reasons for doing so is given below.

1. The plant model is designed once at the beginning and then seldom changed (*e.g.* after a hardware change). The controller design is performed often. Thus the design of a more "benign" plant model can pay off in the long run.

2. In general, for the considered control problem, some restrictions on the class of plant models can be introduced without prejudicing the achievement of the control objectives.

3. Some constraints on the plant class are natural. Also, the need for new algorithms for the computation of more sophisticated controllers can be avoided.

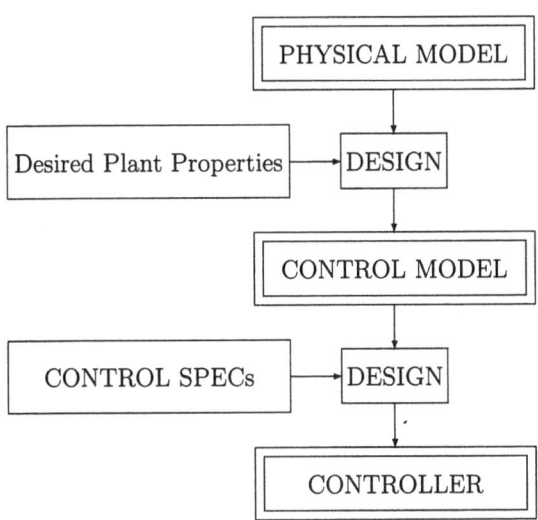

Figure 3: Two-stage design: 1) The control model for the plant is designed from the physical model of the system and the desired plant properties. 2) The controller is designed from the plant model and the control specifications

The design of a plant model can be of great importance for the success of the automatic synthesis of controllers. While the complexity of the computation of a controller for a continuous-valued system depends on the model order, the complexity of the computation of a controller for a discrete event system is not only a function of the size of the model but can heavily depend (and maybe even more crucially) on the structure of the model.

The success of the whole domain of discrete event systems will depend on the future ability of providing the plant models with the necessary *structure* suitable for the controller design.

4.   AN EXAMPLE

We now present as an example the construction procedure for a gas valve of the considered semiconductor manufacturing piece of equipment. The complete alphabets of commands and responses that are modeled are

$$\begin{aligned}
\Sigma_r &= \{\texttt{r\_valve\_failed, r\_valve\_opened, r\_valve\_closed}\} \\
\Sigma_c &= \{\texttt{c\_open\_valve, c\_close\_valve, c\_repair\_valve}\}.
\end{aligned}$$

We adopt the convention that the transition labels starting with "c_" and "r_" denote commands and responses respectively. First, the fundamental process Ξ for the gas valve is modeled by an automaton over the alphabet

$$\Sigma_\Xi = \{\texttt{r\_valve\_opened, r\_valve\_closed, r\_valve\_failed, c\_repair\_valve}\}.$$

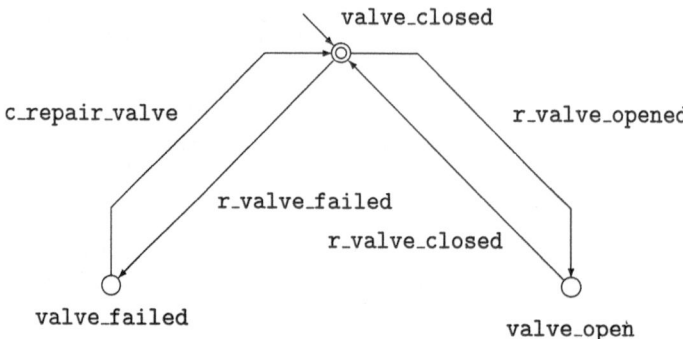

Figure 4: Automaton model of the fundamental valve process Ξ. The arrow on the top indicates the initial state. States marked by two concentric circles are marked states

The automaton is shown in figure 4, where the only marked state is the initial state. The marked language M_Ξ is

$$\{((\texttt{r\_valve\_opened.r\_valve\_closed}) + (\texttt{r\_valve\_failed.c\_repair\_valve}))^*\}$$

and $L_\Xi = \overline{M_\Xi}$. Secondly, the sensing and actuating processes Ψ_i are given by the languages

$$
\begin{aligned}
M_{\Psi_1} &= \{\texttt{c\_open\_valve.}(\texttt{r\_valve\_opened} + \texttt{r\_valve\_failed})\} \\
M_{\Psi_2} &= \{\texttt{c\_close\_valve.r\_valve\_closed}\} \\
M_{\Psi_3} &= \{\texttt{c\_repair\_valve}\}
\end{aligned}
$$

and $L_{\Psi_i} = \overline{M_{\Psi_i}}$. The sequences in M_{Ψ_1} correspond to sensors that indicate whether the command to open the valve is successful or not. The strings in M_{Ψ_2} and M_{Ψ_3} are more detailed descriptions of sequences of events corresponding to closing and repairing the valve. We can see the process Ψ_1 and Ψ_1^* in figure 5.

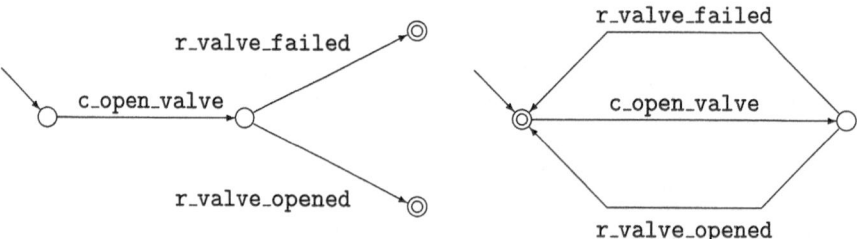

Figure 5: Actuating process Ψ_1 on the left and its infinite repetition Ψ_1^* on the right, described by automata.

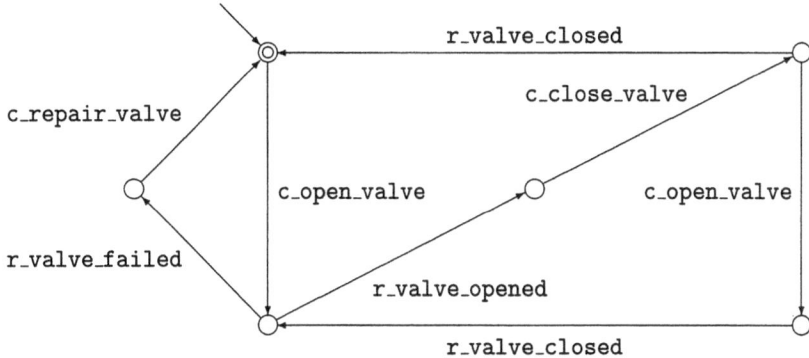

Figure 6: Least restrictive non-blocking plant also allowing the termination of all running routines.

Thirdly, composing the processes as in (6) under constraint equation (7) and of non-blockingness yields the complete input/output plant model of figure 6.

A simpler model for the gas valve process, allowing only one running routine, is given on figure 7.

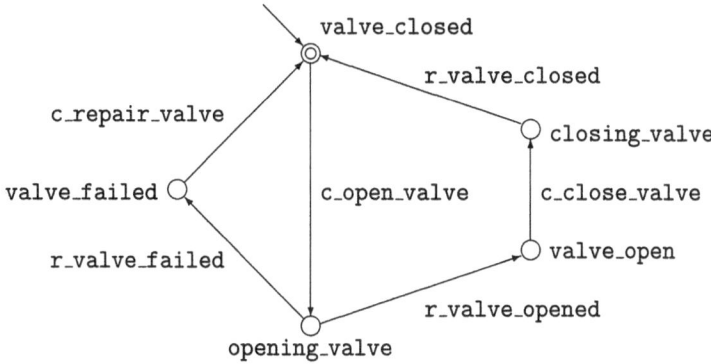

Figure 7: Simple model for the gas valve process P.

It is given by the process $P = (\Sigma, \overline{M_P}, M_P)$ where

$$M_P \doteq \{ \ (\text{c\_open\_valve.}(\text{r\_valve\_opened.c\_close\_valve.r\_valve\_closed} \\ +\text{r\_valve\_failed.c\_repair\_valve}))^* \ \}.$$

Only the initial state of the automaton is marked.

5. CONCLUSION

In this paper an input/output version of the supervisory control problem of Ramadge and Wonham has been introduced. This modified problem is based on an input/output model of both plant and supervisor. Then, the issue of modeling an input/output plant has been addressed, and a methodology for obtaining a plant model has been presented.

The article has tried to show that the modeling of a discrete event system is not an "identification problem" (like in the case of continuous-valued systems) but more a "design problem". During the plant model design, the plant has to be provided with structure which then simplifies the successive controller design problem.

The author believes that the success of the synthesis of controllers for discrete event systems strongly depends on our future ability to introduce structure in our models such as to render computational complexity pheasible in practice.

REFERENCES

[1] P.J. Ramadge and W.M. Wonham. The control of discrete event systems. *Proc. of the IEEE*, 77(1):81–98, January 1989.

[2] C.H. Golaszewski and P.J. Ramadge. Control of discrete event systems with forced events. In *Proc. of 26th Conf. Decision and Control*, pages 247–251, Los Angeles, CA, USA, December 1987.

[3] S. Balemi. A setup for real discrete event system control. Technical Report # 91.07, Automatic Control Laboratory, Swiss Federal Institute of Technology (ETH), Zürich, May 1991.

[4] S. Balemi. Discrete–event systems control of a rapid thermal multiprocessor. In *Proc. of 7th IFAC/IFIP/IFORS/IMACS/ISPE Symposium on Information Control Problems in Manufacturing Technology (INCOM)*, Toronto, Canada, May 1992.

[5] S. Balemi, G.J. Hoffmann, P. Gyugyi, H. Wong-Toi, and G.F. Franklin. Supervisory control of a rapid thermal multiprocessor. *Joint issue Automatica and IEEE Trans. Autom. Control on* Meeting the Challenge of Computer Science in the Industrial Applications of Control, 1993. to appear.

[6] M. Heymann. Concurrency and discrete event control. *IEEE Control System Magazine*, 10(4):103–112, June 1990.

[7] P.J. Ramadge and W.M. Wonham. Supervisory control of a class of discrete event processes. *SIAM J. Control Optim.*, 25(1):206–230, January 1987.

[8] S. Balemi. *Control of Discrete Event Systems: Theory and Application*. PhD thesis, Automatic Control Laboratory, Swiss Federal Institute of Technology (ETH), Zurich, Switzerland, May 1992.

[9] W.M. Wonham and P.J. Ramadge. On the supremal controllable sublanguage of a given language. *SIAM J. Control Optim.*, 25(3):637–659, May 1987.

[10] K. Inan and P. Varaiya. Algebras of discrete event models. *Proc. of the IEEE*, 77(1):24–38, January 1989.

[11] Y. Li and W.M. Wonham. On supervisory control of real–time discrete event systems. *Information Sciences*, 46(3):159–183, 1988.

[12] S. Balemi. Communication delays in connections of input/output discrete event processes. In *Proc. of 31st Conf. Decision and Control*, Tucson, AZ, USA, December 1992.

SUPERVISORY CONTROL AND FORMAL METHODS FOR DISTRIBUTED SYSTEMS

Kemal İnan*

Abstract. A brief introductory exposure for logical discrete event system models is presented. Based on a specific version of this model tailored to supervisory control, some of the mainstream supervisory control problems are formulated in a unified framework.

Formal methods used in software engineering has certain computational and structural similarities to supervisory control and unlike the latter, is closely connected to realistic and widespread practical applications. Formal specification, verification, implementation and testing problems for distributed software are briefly explained. A typical protocol design problem in the context of a layered and peer-leveled architecture is rigorously formulated and discussed. Reasons as to why the simplistic supervisory control formulation - or the *protocol synthesis problem* in the language of protocol engineering - is unrealistic as a design approach are presented. In contrast to protocol design problem, the protocol conversion (gateway) problem is formulated and possible merits of supervisory control approach as a design method are discussed.

1. LOGICAL MODELS FOR DISCRETE EVENT SYSTEMS

A general model from which most of the well-known models can be derived by further structuring is the transition systems model [1]. This model was originally used to capture a system consisting of multiple processors operating in parallel. This model is simply a digraph (finite or infinite) where each node stands for a state and each edge is labeled by an event belonging to an alphabet. Formally if S and Σ stand for the sets of states and events respectively, the graph is characterized by a set of binary relations on S parametrized by events, namely

$$\{R_\alpha \subseteq S \times S | \alpha \in \Sigma\}$$

If each relation R_α is a partial *function* relative to its first argument then the transition system is called *deterministic*. If a certain state is designated as the initial state, then the resulting model is an event-driven state machine.

It is possible to unfold the transition system model at any state by which one obtains a - possibly infinite - rooted tree. Each path starting from the root node in this tree stands for a possible sequence of events corresponding to the labels of the edges of the tree called a *trace* of the system relative to the root. If the original transition system is deterministic then the traces of the unfolded tree completely characterizes the tree since each trace uniquely identifies a node of the tree reached

---

*Electrical & Electronics Engineering Department, Middle East Technical University, Ankara, e-mail: inan@trmetu.bitnet

by the path corresponding to the trace[1]. Therefore a deterministic transition system with an initial state has a *process* representation defined as a prefix-closed subset of Σ^*. If P stands for the process its traces are designated with the symbol trP. Moreover it is not difficult to show that the process (unfolded at the initial state) version of an event driven deterministic state machine corresponds to the minimal (relative to Nerode equivalence) state and reachable version of the state machine in question. The abstract states of the process P are identified by the post-processes P/s, where P/s stands for the subtree of P rooted at the node reached through the trace s.

It is possible to endow states in such a model with attributes for various purposes. For example certain states can be attributed as final states to distinguish selected traces - for example task completion event sequences - in trP. More importantly state attributes can be used to define algebraic operators on processes, that is, if x and y are states in a transition system - that also correspond to processes by unfolding - then a new state $z = x * y$ can be generated if "$*$" is a binary operator on processes.

In the context of supervisory control we define two such attributes on states. Formally we define a process P as a triple $(trP, \alpha P, \tau P)$ where trP is a prefix-closed subset of Σ^* and

$$\alpha P : trP \longrightarrow 2^\Sigma$$

called the *event control set* function and

$$\tau P : trP \longrightarrow \{0, 1\}$$

called the *termination* function.

More generally we define a process P as a pair $(trP, \mu P)$ where

$$\mu P : trP \longrightarrow M$$

is called a *marking* function [2]. For the special case above the marking set is given by $M = 2^\Sigma \times \{0, 1\}$ and the marking function is broken down into its corresponding components, namely the event control and termination functions.

If algebraic operators on processes are used then the transition systems inherit the corresponding algebraic structure. Some of the nodes of a transition system may consist of atomic states say x, y, z, ..., yet some may consist of compound states $x * y$, $x\|z$, ..., using the operators $*$, $\|$, ... If the number of atomic states in a transition system is *finite* then the resulting process relative to the initial state is called a *finitely recursive process* (FRP) [3, 2] over the algebra of operators defined on processes. The expressivity of FRP stems from the fact that one only needs to specify the finite number of transitions from the atomic states and the rest follows from the definition of the operators in a recursive manner. It has been shown in [2] that structured transition systems - such as petri nets for example - can be viewed as FRP with a specific choice of the marking function μ and a marking set M and a specific set of algebraic operators.

Coming back to our specific definition of a process tailored to supervisory control we define a parallel synchronization operator making use of the event

---

[1]Note that this is not the case if the transition system is not deterministic.

control attributes as below. The definition proceeds inductively by first letting

$$\alpha(P\|Q)(<>) := \alpha P(<>) \cup \alpha Q(<>)$$
$$\tau(P\|Q)(<>) := \text{Min}(\tau P(<>), \tau Q(<>))$$

where $<>$ denotes the empty trace, and by defining the one-step progress rule for a possible transition $< a >$ as follows:

$$(P\|Q)/ < a > \begin{cases} := (P/ < a > \|Q/ < a >), \ if \ < a >\in trP \cap trQ \\ := (P/ < a > \|Q), \ if \ < a >\in trP \setminus trQ \ \wedge \ a \notin \alpha Q(<>) \\ := (P\|Q/ < a >), \ if \ < a >\in trQ \setminus trP \ \wedge \ a \notin \alpha P(<>) \\ undefined \ otherwise \end{cases}$$

By using the transitivity of the post-process the definition above can be completed inductively to cover all the traces of $P\|Q$. The parallel operator is associative and commutative. According to the definition above each argument process of the parallel operator can progress individually unless they synchronize on a common transition or one process is blocked by the other one. A simple example to illustrate the use of dynamic event control sets is summarized in the figure below.

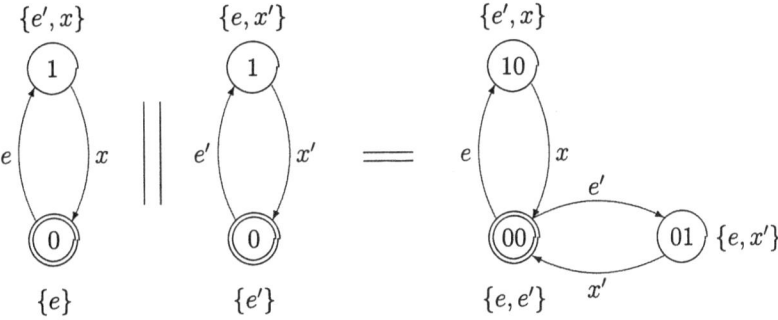

The events e, e' and x, x' stand for entry and exit to the critical section of respective processes and mutual exclusion is accomplished by the choice of event control sets as seen in the figure.

Supervisory Control

We now focus on the processes described in the previous section where the state attributes are event control sets and termination indices. The plant and the supervisor are both modeled as such processes with appropriate constraints. The mechanism of supervision is the parallel operator operating on the plant and the supervisor processes as its arguments.

We first fix the plant process P and assume that its event control function αP is constant and equal to Σ, the universal event alphabet set. This assumption implies that all events originate from the plant.

In contrast, the supervisory process space $Sp = Sp(\Sigma_c, \Sigma_o)$ is the set of processes parametrized by $\Sigma_c, \Sigma_o \subseteq \Sigma$ satisfying the following control and observation constraints:

(1) $\alpha S(s) \subseteq \Sigma_c$, for all $s \in trS$,

(2) All events in trS are in Σ_o,

for any supervisory process S in Sp.

Next we define a partial order "$\leq$" on the set of all processes to structure it into a complete lattice as follows:

$Q \leq R$ iff

(1) $trQ \subseteq R$

(2) $\alpha P(s) = \alpha R(s)$, $\tau Q(s) = \tau R(s)$, for all $s \in trQ$.

Finally we introduce the definition of a *progressive* process as follows: R is a progressive process iff $s \in trR$ implies that there exists a trace t such that $st \in trR$ and $\tau R(st) = 1$. If the termination index stands for task completion instants then this definition simply states that for progressive processes every task can eventually be completed (i.e. no task blocking).

Now let Π denote the following set:

$$\Pi := \{K | K \leq P\}$$

It is easy to verify that Π is a complete lattice under the given partial order. Furthermore let the set of *non-blocking* supervisory processes be defined as follows:

$$Sp' := \{S \in Sp \mid S \| P \text{ is progressive}\}$$

Note that if we let P be a *closed* process, that is $\tau P \equiv 1$, then problems associated with progressivity and non-blocking trivially wither away. This assumption shall be enforced for the so called "interval" problems to be explained below [2].

We now pose supervisory control problems as characterizations of the map:

$$F : Sp'(\Sigma_c, \Sigma_o) \longrightarrow \Pi$$

where $F(S) := P \| S$.

1. The first result in supervisory control [5] is the characterization of the range of F when $\Sigma_o = \Sigma$, that is full observation case. The range turns out to be the set of all progressive processes in Π satisfying the condition

$$(s \in trF(S) \wedge a \notin \Sigma_c) \implies sa \in trF(S)$$

which is defined as the the set of *controllable* processes in Π.

2. Modularity result [6] under full observation assumption simply states that

$$F(S \| S') = F(S) \| F(S')$$

which trivially follows from the commutativity and associativity of "$\|$" and the obvious fact that $P \| P = P$ .In general $Sp'(\Sigma_c, \Sigma)$ is not closed under the operator "$\|$", hence right side above is not necessarily progressive. $F(S)$ and $F(S')$ are called *non-conflicting* (languages) processes, if $F(S) \| F(S')$ is progressive.

---

[2]We have omitted formulations related to marked supervision for the sake of simplicity. Interested reader is referred to [4] for details of process formulation for marked supervision.

3. Results related to the supremal controllable processes [7] is another characterization of F, namely that $F\{Sp'\} \cap \{M | M \leq K \leq P\}$ is a complete upper semi-lattice for any progressive process $K \in \Pi$.

4. Control under partial observations

   Given an observation set Σ_0 the definition of an *observable* process in Π intuitively stands for a progressive process K for which at every state corresponding to a trace $s \in trK$ the information $s.\Sigma_0$ (projection of s on Σ_0) uniquely determines whether $s.a \in trK$ or not for any event a in Σ provided that $s.a \in trP$. The set of observable processes in Π can be shown to be equal to $F\{Sp'(\Sigma, \Sigma_0)\}$. The characterization of the set $F\{Sp'(\Sigma_c, \Sigma_0)\}$, which constitute the progressive process specifications solvable by supervisory control under partial observations was shown by Lin and Wonham [8] to be equal to the set of controllable and observable processes in Π. This involves proving that the image of intersection of two sets is equal to the intersection of the images under the function F as given below

$$F\{Sp'(\Sigma, \Sigma_0) \cap Sp'(\Sigma_c, \Sigma)\} = F\{Sp'(\Sigma, \Sigma_0)\} \cap F\{Sp'(\Sigma_c, \Sigma)\}$$

This is because the right side above is the set of controllable and observable processes and the left side simplifies to $F\{Sp'(\Sigma_c, \Sigma_0)\}$ by using the obvious relation

$$Sp'(\Sigma, \Sigma_0) \cap Sp'(\Sigma_c, \Sigma) = Sp'(\Sigma_c, \Sigma_0)$$

5. Decentralized control with partial observations

   Here the function F has multiple arguments, that is

$$F : \Pi_{i=1}^n Sp(\Sigma_{ci}, \Sigma_{oi}) \longrightarrow \Pi$$

given by

$$F(S_1, \ldots, S_n) := P \| S_1 \| \ldots \| S_n$$

where $S_i \in Sp(\Sigma_{ci}, \Sigma_{oi})$. Cieslak et al. [9] were the first to characterize the range of F. Rudie [10] has introduced the concept of co-observability in an effort to characterize the range as the set of all co-observable and controllable processes in Π. Unfortunately, unlike the centralized partial observation case explained above, the definition of co-observability is coupled to the definition of controllability since it also involves the sets Σ_{ci}. The characterization of the range of F is intuitively simple: at any state corresponding to the trace s at least one supervisor, say the ith one, should have the necessary information $(s.\Sigma_{oi})$ to determine uniquely whether $s.a$ is legal or not, and if not, should also have the authority($a \in \Sigma_{ci}$) to block a - hence the coupling mentioned above.

The problems summarized above, all have their *interval* versions. Let for *closed* processes (all traces marked with index 1!) A and E, an interval $[A, E]$ be defined by the set of processes given by

$$[A, E] := \{K \in \Pi \mid A \leq K \leq E \leq P\}$$

Here the problem is to obtain a condition - preferably necessary and sufficient - in terms of E and A to ensure that the controlled process remains within the interval, that is

$$F\{Sp\} \cap [A, E] \neq \emptyset$$

where we omitted the arguments of the set Sp for simplicity. Crudely put, the results in literature make use of the lattice properties (upper or lower) of the set $F\{Sp\}$ and compute the required supremum and/or infimum processes involving the sets A and/or E.

A problem of considerable practical importance - as to be emphasized in the next section - is the following: find a supervisory process S in $Sp(\Sigma_c', \Sigma_o)$ such that

$$(P\|S).\Sigma_e \in Spec$$

where $Spec$ denotes a set of processes involving only the executive level events belonging to the set Σ_e representing the specification for the controlled process. A simpler version of this problem is when the specification is given as a single process, say R, which yields the requirement

$$(P\|S).\Sigma_e = R$$

The largest process L in Π that satisfies the constraint $L.\Sigma_e = R$ can be shown to be equal to

$$L = P\|R$$

where $\alpha R \equiv \Sigma_e$. However L may not be controllable and/or observable and therefore the problem cannot be reduced to solving $P\|S = L$. The basic problem here, that seems to be bypassed in the supervisory control literature, is the problem of nondeterminism. It is well-known in process semantics that trace (or language) equivalence is not sufficient to depict the behavior of a process. Supervisory control formulations have dealt with nondeterminism by introducing final markings of states representing task completions. In the above context this does not seem to be possible since interpretation (semantics) of $(P\|S).\Sigma_e$ (the projection operation) begs for an explicit definition as a process and trace projection definition is simply not sufficient.

We have merely touched upon the mainstream results and have not mentioned numerous other problems tackled in supervisory control theory. ω-automata and real time extensions [11, 12] , predicate invariance formulations [6],stabilization [13], limited lookahead policies [14],specialized problems on petri nets [15], vector discrete event formulations [17], input/output formulations [16] are among the problems omitted in this brief exposure. This is because our principal purpose is to relate the mainstream problems of supervisory control to another area of active research and application, namely, the area of formal methods and techniques in software engineering. This will be the topic of discussion in the next section.

2. FORMAL METHODS AND SOFTWARE ENGINEERING

The first step in software design and implementation is to be able to capture the requirements as to what the software is supposed to do without getting entangled in implementation details. Once this step is accomplished it is of considerable importance to translate these requirements from an informal schema into a formal - or at least semi-formal - language. A frequent model suggested in software engineering is the waterfall model illustrated in figure 1.

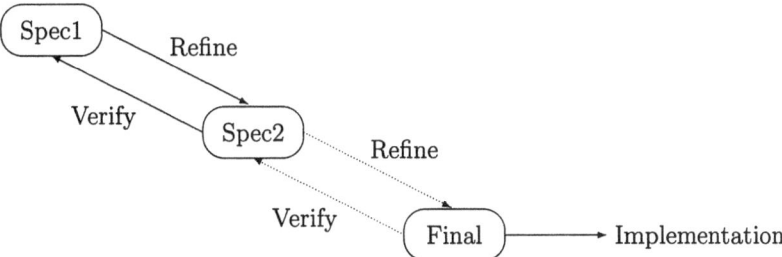

Figure 1: Waterfall Model

Here the steps involved in implementation consist of specification refinement and verification of consistency of the refined version with the previous specification.

There are various uses of such an approach

(1) Efficient client-vendor relation where consensus is achieved at intermediate specifications, before the final implementation.

(2) Interworking of equipment in a heterogeneous environment of technologies and vendors require standardization of interfaces, preferably in a formal manner. Layered OSI (Open Systems Interconnection) model standards issued by CCITT and ISO for communication interface and architectures is a typical example for this requirement. These standards then become the basic guidelines for software implementors on such networks.

(3) The quality of a software produced by such or a similar methodology is necessarily higher than one implemented by a flat approach since intermediate verification steps rule out crucial bugs - at least with high probability ! - that can creep into the final implementation.

(4) Maintaining and configuring the software for later modifications becomes simpler and more systematic once the highest specification layer relevant to the modification is determined. Otherwise a desired local modification may give rise to unintended side effects for the rest of the software.

The above considerations hold, in general, for any kind of software. Our interest, however, lies in problems of *reactive* programming where different program modules are non-terminating and receive, possibly an unbounded string of input data (input stream) and send an output stream of similar structure, all in real time, thereby communicating (synchronizing) with similar modules. Typical of such modules are data communication protocols.

Designing the architecture and implementing and maintaining the software for protocols using formal techniques involves four distinguishable steps.

(1) **Specification**

Services to be supplied by the protocols and the protocols are formally specified. The formalism involved may range from a simple finite state machine to a full-fledge programming language designed specifically for such a purpose. Three such languages have been accepted as CCITT and ISO standards, namely SDL[18], ESTELLE [19] and LOTOS [20]. These languages have been used for formally specifying some of the OSI standard services and protocols [3] .Among these SDL and ESTELLE use infinite buffer queues for synchronization (i.e no blocking of events in synchronization) and are claimed to be based on extended finite state machines - there are implicit algebraic features that surpass extended finite state machine formulations. In contrast LOTOS is algebraically structured, based on a mixture of CSP[23] and CCS [24] and rigid (rendez-vous) synchronization is used.

(2) **Verification**

A formal protocol specification must be verified against the service it is supposed to supply, which, in turn, is formally specified. If both the service and protocol specifications are formally expressed in one of these languages (or a finite state machine for simpler models) then, at least in principle, the formalism involved should allow for automatic verification. In general these languages have verification tools, but for most complex and realistic examples a foolproof verification does not seem to be feasible. In such cases one must be content with partial verification (sometimes referred to as *validation*).

(3) **Implementation**

As illustrated in the waterfall model, it is desirable that the final implementation of a protocol in a specific operating system environment should be derivable , at least in a semi-automatic manner, from the initial formal specification. Much research is done in an effort to automate the generation of the final protocol code via various CASE tools built around a given specification language.

(4) **Conformance Testing**

If a protocol implementation is given as a black box software where only certain input and output primitives are controllable and observable respectively, then conformance testing involves generating test sequences applied to the black box and observing the responses of the black box in order to conclude whether the implementation conforms to the formal standard specification of the protocol. As in verification, this step is rarely fool-proof and theory and formalism is only a partial help.

After this brief exposure to formal methods and their use in protocol software we now formulate the logical problem (as opposed to the performance problem) of

---

[3]The references [21] and especially [22] is informative in explaining why the use of these languages have taken place after the fact, that is, after the standards were already issued.

protocol design in the context of a typical layered architecture. Figure 2 illustrates the problem.

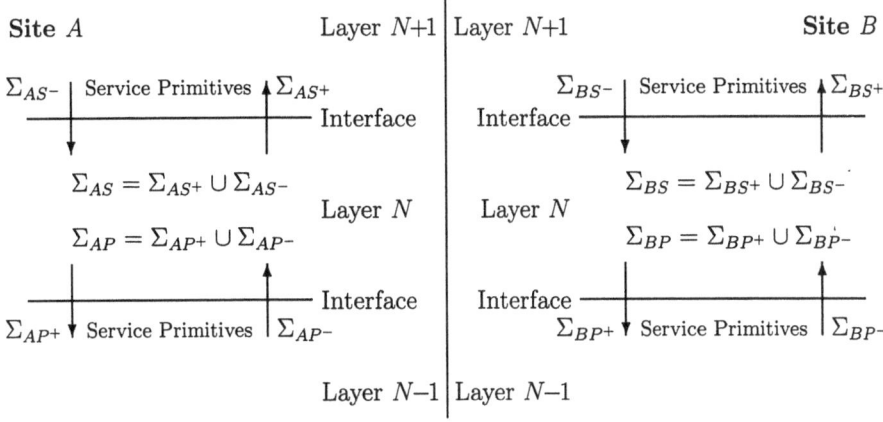

Figure 2: Layered OSI Architecture

Maintaining the definition of a process given in the first section we model both the services and the protocol by processes. In the figure three layers are visible. The protocol to be designed which resides on layer N uses the service supplied by layer $N-1$ for layer N in order to supply the service required by $N+1$. The input and output event sets (set of primitives) as observed from layer N communicating with layers $N+1$ and $N-1$ are indexed by $AS-$, $AP-$ and $AS+$ and $AP+$ respectively and similar conventions hold for the peer level site B as seen in the figure. The definitions Σ_{AP} etc. combine input and output primitive sets at each interface. Define Σ_S and Σ_P as

$$\Sigma_S := \Sigma_{AS} \cup \Sigma_{BS} \; ; \; \Sigma_P := \Sigma_{AP} \cup \Sigma_{BP}$$

which further aggregate layer N and layer $N-1$ service primitives respectively. The formal specification for the service to be supplied by layer N is then a process $QS(\Sigma_S)$ where the argument signifies the events that may occur in process QS. Similarly the formal specification for the service supplied by layer $N-1$ to layer N is a process $QP(\Sigma_P)$. Corresponding *service interfaces* at site A are the *processes* $QS.\Sigma_{AS}$ and $QP.\Sigma_{AP}$ respectively since events at site B are not visible at these interfaces.

The protocol specification problem is to design processes $PA(\Sigma_A)$ and $PB(\Sigma_B)$ respectively at sites A and B such that

$$(PA\|PB\|QP).\Sigma_S = QS$$

Note that processes PA and PB do not directly synchronize and do so indirectly through the process QP. Also observe that under the command of the protocol processes the interface between layers N and $N-1$ at site A (similar situation for site B) gets restricted (or structured) as below

$$(PA\|QP).\Sigma_{AP} \leq QP.\Sigma_{AP}$$

In the above formulation we have not elaborated on the semantics of nondeterminism (equivalence relations used on nondeterministic processes) arising from projections on event sets. Semantic considerations are important, yet beyond the scope of this brief exposure.

There is a good reason why supervisory control formulation - as it stands - cannot be of much help to realistic protocol design problems exemplified above. To formulate the problem as such we may write it as computation of two (decentralized) supervisory processes S_A and S_B such that

$$[QS.\Sigma_{AS}\|S_A\|QP\|S_B\|QS.\Sigma_{BS}].\Sigma_S = QS$$

where the usual control and observation restrictions on S_A and S_B can be formulated in a straightforward manner. The resulting protocols P_A and P_B would then obviously be equal to

$$P_A = QS.\Sigma_{AS}\|S_A \; ; \; P_B = S_B\|QS.\Sigma_{BS}$$

Such a temptation overlooks the actual complexity involved in the problem. In reality events are parametrized and appear as event bundles or groups. For example at layer 4 of OSI (transport services layer) the service supplied by layer 3 is simply the datagram service (i.e. simple transmittal of a packet with arbitrary contents to a given host address with no error checks or sequence ordering). The transport layer protocol structures these packets into different kinds of events (e.g. connection_request, data_request, acknowledgement etc.). Therefore the original bundle of events (packets with arbitrary contents) at layer 3 is highly structured by the protocol to be synthesized in accordance with various needs. These needs usually extend logical concerns and possibly incorporate performance concerns. For example both the alternating bit protocol and a transport protocol using windows for flow control supply the same end-to-end logical service but differ in performance. But once the main skeleton involving all the performance and other concerns of the protocol is given in terms of the specific primitives the rest is a straightforward programming exercise. Reducing this to an on-off natured supervisory control problem is unrealistic except for trivial illustrations of the theory. The formalism that is required, if anything, is a verification tool to compute the projection mentioned above for specified P_A and P_B processes.

There is yet a problem in communication that seems to be suitable for supervisory control formulation: the protocol conversion or the gateway design problem. The reader is referred to [25, 26, 27] for further details and simple examples for the protocol conversion problem in general.

To appreciate why the protocol conversion problem fits supervisory control let us briefly formulate this problem making use of the simple diagram below and the notations and considerations of the previous example.

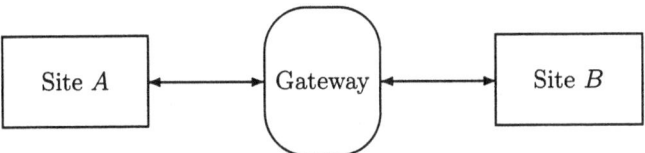

Here we assume that there are two different networks C and C' of the layered structure explained above. The first step is to be able to define a common service between layers N of C at site A and N' of C' at site B'. Typically this could be a least common denominator of their services (if one supplies the transport service and the other the datagram service then the common denominator is the datagram service). In general, this step involves explicitly defining a process

$$QS'' = QS''(\Sigma_{AS}, \Sigma_{B'S'})$$

as the common service to be jointly supplied by layers N and N' in C and C'. At the sites A and B' the corresponding interfaces should not be able to distinguish this from the services QS and QS' they supply (interface fooling conditions) since we are not allowed to change the protocols or primitives for C and C'. More precisely

$$QS''.\Sigma_{AS} \leq QS.\Sigma_{AS} \; ; \; QS''.\Sigma_{B'S'} = QS'.\Sigma_{B'S'}$$

The problem is to find a supervisor S such that

$$(P_A\|QP\|P_B\|S\|P_{A'}\|QP'\|P_{B'}).\{\Sigma_{AS} \cup \Sigma_{B'S'}\} = QS''$$

The formulation for controllable and observable events readily follows from geographic separation conditions. The actual gateway G is then equal to

$$G = P_B\|S\|P_{A'}$$

Note that this formulation also guarantees the lower level interface fooling conditions using the properties of projection (not elaborated in this paper) since

$$(P_A\|QP\|G).\Sigma_{AP} \;\leq\; (P_A\|QP\|P_B).\Sigma_{AP}$$
$$(P_{B'}\|QP'\|G).\Sigma_{B'P'} \;\leq\; ((P_{B'}\|QP'\|P_{A'}).\Sigma_{B'P'}$$

The reason why this problem fits nicely into the supervisory control theory is simply because no new primitives are required for the conversion problem: conversion requires sequential interleaving coordination of primitives in P_B and $P_{A'}$. Yet this is just an initial optimistic consideration and further research is required to relate a realistic gateway design problem to a possible supervisory control formulation.

REFERENCES

[1] R. Keller. Formal verification of parallel programs. *Communications of the ACM*, 19(7):371–384, 1976.

[2] K. Inan and P. Varaiya. Algebras of discrete event models. *IEEE Proceedings*, 77(1):24–38, January 1989.

[3] K. Inan and P. Varaiya. Finitely recursive process models for discrete event systems. *IEEE Trans. Automatic Control*, 33(7):626–639, July 1988.

[4] K. Inan. An algebraic approach to supervisory control. *MCSS*, (5):151–164, 1992.

[5] P.R. Ramadge and W.M. Wonham. Supervisory control of a class of discrete-event processes. *SIAM Journal on Control and Optimization*, 25(1):206–230, January 1987.

[6] P.R. Ramadge and W.M. Wonham. Modular feedback logic for discrete-event systems. *SIAM Journal on Control and Optimization*, 25(5):1202–1218, May 1987.

[7] P.R. Ramadge and W.M. Wonham. On the supremal controllable sublanguage of a given language. *SIAM Journal on Control and Optimization*, 25:637–659, May 1987.

[8] F. Lin and W.M. Wonham. On observability of discrete-event systems. *Information Sciences*, 44:173–198, 1988.

[9] R. Cieslak, C. Desclaux, A. Fawaz, and P. Varaiya. Supervisory control of discrete event processes with partial observations. *IEEE Trans. Automatic Control*, 33(3):249–260, March 1988.

[10] K. Rudie. *Decentralized Control of Discrete Event Systems*. PhD thesis, University of Toronto, 1992.

[11] J.G. Thistle and W.M. Wonham. Control problems in temporal logic framework. *International Journal on Control*, 44(4):943–976, May 1986.

[12] J.S. Ostroff and W.M. Wonham. Real time computer control of discrete event systems modeled by extended state machines. Technical Report Rep.8618, Dept. Electrical Engineering, University of Toronto, September 1986.

[13] C. M. Ozveren and A. S. Willsky. Output stabilizability of discrete event dynamic systems. *IEEE Trans. Automatic Control*, AC-36(8):925–935, August 1991.

[14] S. L. Chung, S. Lafortune, and F. Lin. Limited lookahead policies in supervisory control of discrete event systems. Technical report, University of Michigan, 1991.

[15] L. E. Holloway and B. H. Krogh. Synthesis of feedback control logic for a class of controlled petri nets. *IEEE Trans. Automatic Control*, AC-356(5):514–523, May 1990.

[16] S. Balemi. *Control of Discrete Event Systems: Theory and Application.* PhD thesis, Swiss Federal Institute of Technology (ETH), Zürich, May 1992.

[17] Y. Li. *Control of Vector Discrete Event Systems.* PhD thesis, University of Toronto, July 1991.

[18] F. Belina and D. Hogrefe. Introduction to sdl. In K. J. Turner, editor, *FORTE I.* North Holland, 1989.

[19] M. Diaz et al., editors. *The Formal Description Technique Estelle.* North Holland, 1989.

[20] P. van Eijk et al., editors. *The Formal Description Technique LOTOS.* North Holland, 1989.

[21] Gregor von Bochmann. Formal methods for describing distributed systems: A discussion of the experience in osi standardization. Technical Report 712, University of Montreal, January 1990.

[22] C.A. Vissers. FDTs for open distributed systems, a retrospective and a prospective view. In L. Logrippo, R. L. Probert, and H. Ural, editors, *Protocol Specification, Testing and Verification, X,* pages 341–362, Amsterdam, June 1990. IFIP, North-Holland.

[23] C.A.R Hoare. *Communicating Sequential Processes.* Prentice-Hall International, Herts, England, 1985.

[24] R. Milner. *Calculus of Communicating Systems.* Springer, New York, NY, 1980.

[25] P.E. Green. Protocol conversion. *IEEE Trans. Communication,* COM-34(3):257–268, March 1986.

[26] S.S. Lam. Protocol conversion. *IEEE Trans. Software Engineering,* SE-14(3):353–362, March 1988.

[27] G.V. Bochmann. Deriving protocol converters for communications gateways. *IEEE Trans. Communication,* COM-38(9):1298–1300, September 1990.

AN OVERVIEW OF RESULTS IN DISCRETE EVENT SYSTEMS USING A TRACE THEORY BASED SETTING

Rein Smedinga[*]

Abstract. Discrete event systems can be modelled using a trace theory based setting. A control problem similar to the one in supervisory control theory can be formulated and easily solved using a special operator, the reflection. Using state graphs the solution can be computed effectively. Moreover, a subsystem of the controller can be computed such that the connection of plant and controller is free of lock.

1. INTRODUCTION

Different approaches and notation exist in the field of logical discrete event systems. Here, we try to give the most simple and clear notation that is possible, without loosing expressive power. We use the trace theory, developed in 1984. Trace theory was developed for VLSI-design [1], but seems also very suitable in modelling more general discrete event systems.

In trace theory symbols are used to model the fact that there exists some signal on some wire: the event a means that wire a (sometimes called channel) contains a message. We can extend this idea to work on more general discrete event systems in the following way: suppose we have some observer who looks at some system from a distance. He only has pen and paper and writes down a symbol on that paper each time he sees a corresponding event occurring in the system. At some time there will be a string of symbols on his paper. This sequence tells him what events have occurred and in what order. If the contents of the paper is abc it means that first event a occurred, then event b and, last, event c. Systems do not necessarily behave in precisely one way, which means that the observer might also have seen some other events or order of events to happen. The system might also be in rest, i.e., do nothing at all, in which case the observer will have an empty piece of paper (which we denote by ϵ).

Now, we define *the* behaviour of a system as all possible sequences of events an observer of the system could have written down.

As an extension to trace theory we add that an observer also has the possibility to see if a system has completed a task. Sequences that correspond to such a completed task are collected in a separate set, called the task set of the system.

Because the observer only has one pen he cannot write down two events at the same time. This means that two events cannot occur at the same time. If

[*]Department of computing science, University of Groningen, P.O.Box 800, NL-9700 AV Groningen, the Netherlands, e-mail: rein@cs.rug.nl

in some system two events (say a and b) can occur in parallel, it is even likely that the observer writes down on his paper a followed by b as b followed by a. If the behaviour of some systems contains sequences $xaby$ and $xbay$ (where x and y are arbitrary sequences) we have no possibility to distinguish between parallel occurrence of a and b and choice between ab and ba. This difference cannot be modelled in normal language-based approaches. Only Petri Nets have this possibility.

In trace theory, connection of systems is done by connecting some wires. A connected system now contains connected wires (which will be named the same in both systems) and some unconnected wires. A signal on a connected wire in the first system automatically means that the same signal is on the corresponding wire of the second system. Or, to say it a little bit different: a common event in one system can occur only if the same event in the other system can occur as well. This idea will also be extended to more general discrete event systems. An event that is common to both connected systems only occurs if both systems can engage in it. So events only occur at the same time, if they are in fact the same event.

Connecting systems in this way is defined more often in logical DES-modelling and known under the name of synchronisation.

Common events in a connection are sometimes considered as internal events of the connection, not visible by the observer. In that case we have the so called external connection (i.e., a connection where internal events are no longer visible and will not be written down by the observer).

Example 1

$$P = \langle \{in, out\},$$
$$\{\epsilon, \ in, \ in \cdot out, \ in \cdot out \cdot in, \ldots\},$$
$$\{\epsilon, \ in \cdot out, \ in \cdot out \cdot in \cdot out, \ldots\}\rangle$$

is a one-place buffer (A $\cdot$ denotes concatenation here; $\cdot$ is only used when the events are denoted by names. It is omitted when the events are denoted by simple symbols). Possible events are in and out; each alternating sequence of events in and out with at least as many in's as out's is a behaviour and each alternating sequence of in's and out's with as many in's as out's is a completed task.   □

2.   DEFINITIONS

According to our intuitive ideas we have the following definition of a discrete event system (see [2, 3, 4]):

Definition 1 *A Discrete event logical system (DES) is defined as a triple*

$$P = \langle \mathbf{a}P, \mathbf{b}P, \mathbf{t}P \rangle$$

where:
$\mathbf{a}P$ *is the alphabet (a finite set of symbols, representing the events),*

$\mathbf{b}P \subseteq (\mathbf{a}P)^*$ *the behaviour set*
$\mathbf{t}P \subseteq (\mathbf{a}P)^*$ *the task set*
Both $\mathbf{b}P$ *and* $\mathbf{t}P$ *are trace sets, i.e., possibly infinite, set of strings over the alphabet.*

In order to obtain a realistic system we add the following restrictions to the definition:[1]

$$\mathbf{pref}(\mathbf{b}P) = \mathbf{b}P \qquad \mathbf{t}P \subseteq \mathbf{b}P$$

(i.e., the behaviour is prefix closed and each task is a behaviour). Such a system is called a *realistic system*. Unrealistic systems go beyond our intuitive idea of a DES (how can an observer find a sequence abc on his paper without ever having the sequence ab on it). Nevertheless, unrealistic systems play a crucial role in finding a controller, as we shall see later.

Example 1 (cont.) We omit the behaviour set in the description of some system if it is obvious from the task set (i.e., $\mathbf{b}P = \mathbf{pref}(\mathbf{t}P)$). Moreover, more or less standard notation from regular languages exists to denote behaviour and task set in a somewhat shorter and clearer way. P written in this way gives:

$$P = \langle \{in, out\}, (in \cdot out)^* \rangle$$

where * stands for the Kleene-star operator. Another useful operator is $|$, denoting choice, e.g., $(a(b|c))^*$ stands for

$$\{\epsilon, a, ab, ac, aba, aca, abab, abac, acab, acac, \ldots\}$$

It should be emphasized that the use of some operators, normally used for regular languages, does not imply that the behaviour and task sets of a system must be regular. In fact, we do not impose any restriction on these sets!   □

Some systems have been given a special name: The empty system $\mathbf{empty}(A) = \langle A, \emptyset, \emptyset \rangle$ has no behaviour at all. It stands for the impossible system. The skip system $\mathbf{skip}(A) = \langle A, \{\epsilon\}, \{\epsilon\} \rangle$ models a system in rest (only able to do nothing) and the chaos system $\mathbf{chaos}(A) = \langle A, A^*, A^* \rangle$ models a system that can do every combination of events.

In a connection of two systems common events can only occur if both systems can engage in it. An other way of formalising our intuitive idea of connecting systems is that each behaviour (and task set) in the connection must be such that a sequence, restricted to the alphabet of one of the systems, is an element of the behaviour (task set, respectively) of that system:

Definition 2 *The connection of two systems* P *and* R *is defined by*

$$P \parallel R = \langle \mathbf{a}P \cup \mathbf{a}R, \{x \mid x \in (\mathbf{a}P \cup \mathbf{a}R)^* \wedge x \lceil \mathbf{a}P \in \mathbf{b}P \wedge x \lceil \mathbf{a}R \in \mathbf{b}R\},$$
$$\{x \mid x \in (\mathbf{a}P \cup \mathbf{a}R)^* \wedge x \lceil \mathbf{a}P \in \mathbf{t}P \wedge x \lceil \mathbf{a}R \in \mathbf{t}R\} \rangle$$

$\lceil$ stands for alphabet restriction: $x \lceil A$ is the projection of string x on alphabet A, i.e., all symbols not in A are removed from x.

---

[1]The prefix closure of a trace set is defined by $\mathbf{pref}(T) = \{x \mid (\exists u :: xu \in T)\}$.

Example 2 Consider $P = \langle \{a, b\}, (ab) \rangle$ and $R = \langle \{b, c\}, (cb) \rangle$, then

$$P \parallel R = \langle \{a, b, c\}, ((ac|ca)b) \rangle$$

For example, $ac \in b(P \parallel R)$ because $ac\lceil aP = a \in bP$ and $ac\lceil aR = c \in bR$. $\quad\square$

Property 1

(a) *symmetry:* $\quad P \parallel R = R \parallel P$
(b) *unit element:* $\quad P \parallel \mathbf{skip}(\emptyset) = P$
(c) *zero element:* $\quad P \parallel \mathbf{empty}(\emptyset) = \mathbf{empty}(aP)$
(d) *idempotenty:* $\quad P \parallel P = P$
(e) *associativity:* $\quad (P \parallel R) \parallel S = P \parallel (R \parallel S)$

If common events are not important we can use the external connection. The observer is now unable to see the common events because they are internal w.r.t. the connection.

Definition 3 *The external connection is defined by*

$$P \rceil\!\lceil R = (P \parallel R)\lceil(aP \div aR)$$

(where $\quad A \div B = (A \cup B) \setminus (A \cap B)$ is the symmetric set difference and $\quad \langle A, B, T \rangle \lceil A_1 = \langle A \cap A_1, B\lceil A_1, T\lceil A_1 \rangle$.)

Example 2 (cont.) Reconsider P and R. It is easily verified that $P \rceil\!\lceil R = \langle \{a, c\}, (ac|ca) \rangle$. $\quad\square$

Property 2

(a) *symmetry:* $\quad P \rceil\!\lceil R = R \rceil\!\lceil P$
(b) *unit element:* $\quad P \rceil\!\lceil \mathbf{skip}(\emptyset) = P$
(c) *zero element:* $\quad P \rceil\!\lceil \mathbf{empty}(\emptyset) = \mathbf{empty}(aP)$
(d) *non-idempotenty:* $\quad P \neq \mathbf{empty}(\emptyset) \Rightarrow P \rceil\!\lceil P = \mathbf{skip}(\emptyset)$

$\rceil\!\lceil$ is, in general, not associative. However, we have:

Property 3 *If no event occurs in more than two of the alphabets, the operator $\rceil\!\lceil$ is associative.*

Systems can be ordered. We say system P is a subsystem of system R, if everything that P can do also can be done by R.

Definition 4 *P is a subsystem of R, notation $P \subseteq R$, if*

$$aP = aR \wedge bP \subseteq bR \wedge tP \subseteq tR$$

Notice that we only order systems with equal alphabets.

Property 4

(a) $\quad R_1 \subseteq R_2 \Rightarrow (P \parallel R_1) \subseteq (P \parallel R_2)$
(b) $\quad R_1 \subseteq R_2 \Rightarrow (P \rceil\!\lceil R_1) \subseteq (P \rceil\!\lceil R_2)$

Example 3 We have (for arbitrary alphabet A):

$$\mathbf{empty}(A) \subseteq \mathbf{skip}(A) \subseteq \mathbf{chaos}(A)$$

$\mathbf{empty}(A)$ is the smallest system with alphabet A, $\mathbf{chaos}(A)$ is the largest. It can be shown that the class of all discrete event systems over some alphabet A is a complete lattice. $\quad\square$

3.   A CONTROL PROBLEM

In our trace theory based setting, systems will be considered having two kinds of events: controllable and uncontrollable events. Such systems can be controlled (what will be the goal of the control will be explained later) using these controllable events. The alphabet of the controller will consists of (part of) the controllable events of the system. Controllable events therefore may be common to both system and controller and used as internal events in the connection of system and controller. Uncontrollable events remain visible in an external connection and should be controlled, using the controllable events, to behave in some predescribed manner.

Controllable events are sometimes considered as commands and the uncontrollable events as responses ([5]). Controlling a system can thus be considered as giving the right commands in order for the system to answer with the desired responses. One way of defining such a control problem is giving a minimal desired and a maximal desired response-behaviour and try to find a controller, such that the connection of system and controller indeed has response-behaviour between the given limits.

In the following, we assume that a controller can use all the possible commands of the system and that the restrictions on the resulting connection are given for all possible responses.[2] Now we can formulate our control problem, in a way similar to the control problem in supervisory theory, see [6, 7].

Control problem 1 (CODE) *Given are a DES P with control-events $cP \subseteq aP$ and two DESs $L_{min} \subseteq L_{max} \subseteq$* **chaos**$(eP)$ *(where $eP = aP \setminus cP$). To be found a controller system R with $aR = cP$ such that*

$$L_{min} \subseteq P \,]\!\lceil \, R \subseteq L_{max} \qquad \text{(minmax condition)}$$

Notice that $P \,]\!\lceil \, R = (P \parallel R)\lceil eP$, so minmax condition restricts the response behaviour and response tasks of the controlled system to be within certain limits.

A first guess for the controller might be to compute $P \,]\!\lceil \, L$ for some L with $L_{min} \subseteq L \subseteq L_{max}$. However, it can be proven ([4]) that

$$(P \,]\!\lceil \, L) \,]\!\lceil \, L \supseteq L$$

so, even if we start with $L = L_{min}$ we might end in an external connection that is too large. However, it turns out that complementary systems may lead to a solution of our control problem. Instead of computing $P \,]\!\lceil \, L$ we compute the external connection of P and the complement of L and afterwards, compute the complement of this connection. To make this idea more formal, we first introduce the complement of a system.

4.   THE REFLECTION OPERATOR

Definition 5 *The reflection of a system P, denoted $\sim P$, is defined by*

$$\langle aP, (aP)^* \setminus bP, (aP)^* \setminus tP \rangle$$

---

[2]This is without loss of generality: if not all commands may be used, temporally change the unused commands into responses. If the goal of the control does not affect all responses, simply add to the limits all possibilities of these responses.

Notice that if P is realistic, $\sim P$ need not be. $\sim P$ may contain behaviour that is not prefix closed and/or tasks that are no behaviour.

The reflection operator was first introduced in [8] as the greatest testing environment of a system. In our trace theory based setting this operator turns out to be just the complementary system.

Example 4 Consider $P = \langle \{a\}, \{\epsilon, a, aa\}, \{aa\} \rangle$, then

$$\sim P = \langle \{a\}, \{aaa, aaaa, \cdots\}, \{\epsilon, a, aaa, aaaa, \cdots\} \rangle$$

Notice that the behaviour of $\sim P$ is no longer prefix closed (e.g. ϵ is no behaviour). Moreover, not every task is a behaviour (e.g. ϵ and a are no behaviour).   □

Property 5
(a) $P \subseteq R \Leftrightarrow \sim R \subseteq \sim P$
(b) $\sim\sim P = P$
(c) $\sim\mathbf{skip}(\emptyset) = \mathbf{empty}(\emptyset)$
(d) $\sim\mathbf{chaos}(A) = \mathbf{empty}(A)$

5.   SOLUTION FOR CODE

We claim ([3, 4]) that
$$F(P, L) = \sim(P \,]\!\lceil\, \sim L)$$

leads to a solution. To prove the main result, we first need two more properties.
Property 6

$$P \subseteq R \Leftrightarrow P \,]\!\lceil\, \sim R = \mathbf{empty}(\emptyset)$$

Proof:
$\quad P \subseteq R$
$\Leftrightarrow \qquad [\text{ definition of } \subseteq]$
$\quad \mathbf{a}P = \mathbf{a}R \wedge \mathbf{b}P \subseteq \mathbf{b}R \wedge \mathbf{t}P \subseteq \mathbf{t}R$
$\Leftrightarrow \qquad [\text{ set theory }]$
$\quad \mathbf{a}P = \mathbf{a}R \wedge \mathbf{b}P \cap ((\mathbf{a}R)^* \setminus \mathbf{b}R) = \emptyset \wedge \mathbf{t}P \cap ((\mathbf{a}R)^* \setminus \mathbf{t}R) = \emptyset$
$\Leftrightarrow \qquad [\text{ definition of } \sim]$
$\quad \mathbf{a}P = \mathbf{a}(\sim R) \wedge \mathbf{b}P \cap \mathbf{b}(\sim R) = \emptyset \wedge \mathbf{t}P \cap \mathbf{t}(\sim R) = \emptyset$
$\Leftrightarrow \qquad [\text{ definition of }]\!\lceil\,]$
$\quad P \,]\!\lceil\, \sim R = \langle \emptyset, \emptyset, \emptyset \rangle$

Property 7

$$P \,]\!\lceil\, R \subseteq S \Leftrightarrow R \subseteq F(P, S)$$

Proof:
$\quad P \,]\!\lceil\, R \subseteq S$
$\Leftrightarrow \qquad [\,]\!\lceil\, \text{ is symmetric }]$
$\quad R \,]\!\lceil\, P \subseteq S$
$\Leftrightarrow \qquad [\text{ property 6 and definition of } \subseteq]$
$\quad (R \,]\!\lceil\, P) \,]\!\lceil\, \sim S = \mathbf{empty}(\emptyset) \wedge \mathbf{a}P \div \mathbf{a}R = \mathbf{a}S$
$\Leftrightarrow \qquad [\, \mathbf{a}P \cap \mathbf{a}R \cap \mathbf{a}S = \emptyset \Rightarrow]\!\lceil\, \text{ is associative }]$

$$R \rceil\!\lceil (P \rceil\!\lceil \sim\!S) = \mathbf{empty}(\emptyset) \wedge \mathbf{a}P \div \mathbf{a}R = \mathbf{a}S$$
$$\Leftrightarrow \qquad [\text{ property 5 (b) and set theory }]$$
$$R \rceil\!\lceil \sim\!\sim\!(P \rceil\!\lceil \sim\!S) = \mathbf{empty}(\emptyset) \wedge \mathbf{a}R = \mathbf{a}P \div \mathbf{a}S$$
$$\Leftrightarrow \qquad [\text{ property 6 and definition of } \subseteq]$$
$$R \subseteq \sim\!(P \rceil\!\lceil \sim\!S)$$

Theorem 1 *The control problem has a solution if and only if*

$$L_{min} \subseteq G(P, L_{max})$$

and, if it is solvable, the greatest solution (with respect to $\subseteq$) is $F(P, L_{max})$.
Proof:

$$(\exists R :: L_{min} \subseteq P \rceil\!\lceil R \subseteq L_{max})$$
$$\Leftrightarrow \qquad [\text{ property 7 }]$$
$$(\exists R :: L_{min} \subseteq P \rceil\!\lceil R \wedge R \subseteq F(P, L_{max}))$$
$$\Leftrightarrow \qquad [\Rightarrow: \rceil\!\lceil \text{ is monotonic}, \Leftarrow: \text{ take } R = F(P, L_{max})]$$
$$L_{min} \subseteq P \rceil\!\lceil F(P, L_{max})$$

It is straightforward to see that $F(P, L_{max})$ is indeed the greatest possible solution.

Example 5 Consider $P = \langle \{a, b\}, (ab) \rangle$ and $L = L_{min} = L_{max} = \langle \{b\}, (b) \rangle$. Computing the candidate controller leads to:

$$\begin{aligned} \sim\!L &= \langle \{b\}, \{bb, bbb, \ldots\}, \{\epsilon, bb, bbb, \ldots\} \rangle \\ P \parallel \sim\!L &= \mathbf{empty}(\{a, b\}) \\ P \rceil\!\lceil \sim\!L &= \mathbf{empty}(\{a\}) \\ \sim\!(P \rceil\!\lceil \sim\!L) &= \mathbf{chaos}(\{a\}) \end{aligned}$$

So $F(P, L) = \mathbf{chaos}(\{a\})$. This controller is minimal restrictive. Moreover, it also contains all traces that do not have any effect on the system P. For example the behaviour aaa in the controller does not have any effect on P, because P can not engage in that behaviour. The controller also has ϵ as possible task, but because P does not start with a completed task, this has no effect either. $\quad\square$

Property 8
(a) $L_1 \subseteq L_2 \Rightarrow F(P, L_1) \subseteq F(P, L_2)$
(b) $R \subseteq F(P, P \rceil\!\lceil R)$
(c) $P \rceil\!\lceil F(P, P \rceil\!\lceil R) = P \rceil\!\lceil R$
Proof:
(a) $\qquad L_1 \subseteq L_2$
$$\Leftrightarrow \qquad [\text{ property 5 (a) }]$$
$$\sim\!L_1 \supseteq \sim\!L_2$$
$$\Rightarrow \qquad [\text{ property 4 (b) }]$$
$$P \rceil\!\lceil \sim\!L_1 \supseteq P \rceil\!\lceil \sim\!L_2$$
$$\Leftrightarrow \qquad [\text{ property 5 (a) }]$$
$$\sim\!(P \rceil\!\lceil \sim\!L_1) \subseteq \sim\!(P \rceil\!\lceil \sim\!L_2)$$
$$\Leftrightarrow \qquad [\text{ definition of } F]$$
$$F(P, L_1) \subseteq F(P, L_2)$$

(b) follows from property 7 if we take $S = P \,][\, R$.

For (c) we prove that $P \,][\, F(P, P \,][\, R) \subseteq P \,][\, R$ because $\supseteq$ follows from (a). We show that the task set of the lhs. is included in the rhs.

$x \in \mathbf{t}(P \,][\, F(P \,][\, R))$

$\Leftrightarrow$ [definition of $][$]

$(\exists y : y \in \mathbf{t}P \wedge y\lceil \mathbf{c}P \in \mathbf{t}(F(P, P \,][\, R)) : y\lceil \mathbf{e}P = x)$

$\Leftrightarrow$ [definition of F and $\sim$ and $y \in \mathbf{t}P \Rightarrow y\lceil \mathbf{c}P \in (\mathbf{e}P)^*$]

$(\exists y : y \in \mathbf{t}P \wedge y\lceil \mathbf{c}P \notin \mathbf{t}(P \,][\, \sim(P \,][\, R)) : y\lceil \mathbf{e}P = x)$

$\Leftrightarrow$ [definition of $][$]

$(\exists y : y \in \mathbf{t}P \wedge (\forall z : z \in \mathbf{t}P \wedge$
$\qquad\qquad z\lceil \mathbf{e}P \in \mathbf{t}(\sim(P \,][\, R)) : z\lceil \mathbf{c}P \neq y\lceil \mathbf{c}P) : y\lceil \mathbf{e}P = x)$

$\Leftrightarrow$ [definition of $\sim$ and $z \in \mathbf{t}P \Rightarrow z\lceil \mathbf{e}P \in (\mathbf{e}P)^*$]

$(\exists y : y \in \mathbf{t}P \wedge (\forall z : z \in \mathbf{t}P \wedge$
$\qquad\qquad z\lceil \mathbf{e}P \notin \mathbf{t}(P \,][\, R) : z\lceil \mathbf{c}P \neq y\lceil \mathbf{c}P) : y\lceil \mathbf{e}P = x)$

$\Leftrightarrow$ [definition of $][$]

$(\exists y: y \in \mathbf{t}P \wedge (\forall z : z \in \mathbf{t}P \wedge (\forall u : u \in \mathbf{t}P \wedge u\lceil \mathbf{c}P \in \mathbf{t}R : u\lceil \mathbf{e}P \neq z\lceil \mathbf{e}P)$
$\qquad\qquad : z\lceil \mathbf{c}P \neq y\lceil \mathbf{c}P)$
$\qquad : y\lceil \mathbf{e}P = x)$

$\Leftrightarrow$ [trading]

$(\exists y : y \in \mathbf{t}P \wedge (\forall z : z \in \mathbf{t}P \wedge z\lceil \mathbf{c}P = y\lceil \mathbf{c}P$
$\qquad\qquad : (\exists u : u \in \mathbf{t}P \wedge u\lceil \mathbf{c}P \in \mathbf{t}R : u\lceil \mathbf{e}P = z\lceil \mathbf{e}P))$
$\qquad : y\lceil \mathbf{e}P = x)$

$\Rightarrow$ [take $z = y$]

$(\exists y : y \in \mathbf{t}P \wedge (\exists u : u \in \mathbf{t}P \wedge u\lceil \mathbf{c}P \in \mathbf{t}R : u\lceil \mathbf{e}P = y\lceil \mathbf{e}P)) : y\lceil \mathbf{e}P = x)$

$\Rightarrow$

$(\exists u : u \in \mathbf{t}P \wedge u\lceil \mathbf{c}P \in \mathbf{t}R : u\lceil \mathbf{e}P = x)$

$\Leftrightarrow$

$x \in \mathbf{t}(P \,][\, R)$

Similar for the behaviour set, which completes the proof.

Because of using the reflection operator twice, the resulting controller need not be realistic any more. In order to find a realistic controller we must compute the so called des-interior of $F(P, L)$, where the des-interior of P is defined by

$$\mathbf{des}(P) = \langle \mathbf{a}P, \{x \mid x \in \mathbf{b}P \wedge (\forall z : z \in \mathbf{pref}(x) : z \in \mathbf{b}P)\},$$
$$\{x \mid x \in \mathbf{t}P \wedge (\forall z : z \in \mathbf{pref}(x) : z \in \mathbf{b}P)\}\rangle$$

Example 6 Consider $P = \langle \{a, b\}, \{\epsilon, a, aab, b, ab\}, \{ab, aab, ba\}\rangle$. Then we have:
$\mathbf{des}(P) = \langle \{a, b\}, \{\epsilon, a, b, ab\}, \{ab\}\rangle = \langle \{a, b\}, (ab)\rangle$. $\qquad\qquad\square$

Because the des-interior of some system P is the greatest realistic subsystem of P we have:

Theorem 2 *The control problem is solvable with a realistic solution if and only if*

$$L_{min} \subseteq P \,][\, \mathbf{des}(F(P, L_{max}))$$

and the greatest realistic solution then is $\mathbf{des}(F(P, L_{max}))$.

6. OBSERVABILITY

Next, we would like to investigate if some smallest solution for our CODE-problem can be found. Because every solution of CODE can be extended with tasks that have no influence on the result (take $R_{new} = R \cup R_{ext}$ for an R_{ext} with $P \parallel R_{ext} = \mathbf{empty}(\mathbf{a}P)$, then R_{new} is also a solution), and also, from every solution such tasks can be deleted, we can only wish that within the class of solutions R with $R \subseteq P\lceil \mathbf{c}P$ some smallest one can be found.

Therefore, we introduce the notion *effective part* of a solution R, which is equal to $R \cap P\lceil \mathbf{c}P$, where $P \cap R$ is defined only for systems with equal alphabets as $P \parallel R$. $P \cap R$ contains behaviour (tasks) that are both in the behaviour set of P and R (task sets, respectively), i.e.,

$$P \cap R = \langle \mathbf{a}P, \mathbf{b}P \cap \mathbf{b}R, \mathbf{t}P \cap \mathbf{t}R \rangle$$

Now, suppose R is a solution of CODE, then $P \rceil\lceil R$ satisfies the minmax condition. From property 8 (b) we have that $R \subseteq F(P, P \rceil\lceil R)$. In general, we have no equality here, as is shown in the following example.

Example 7 Consider $P = \langle \{a, b, d, e\}, (ad(be|ae)) \rangle$ with $\mathbf{e}P = \{d, e\}$ and $L_{min} = L_{max} = \langle \{d, e\}, (de) \rangle$. We can verify that $R = \langle \{a, b\}, (aa) \rangle$ is a solution of CODE, but no L satisfying the minmax condition can be found so that $F(P, L) \cap P\lceil \mathbf{c}P = R$. The only possible L, namely $L = L_{min}$, leads to $F(P, L) \cap P\lceil \mathbf{c}P = \langle \{a, b\}, (a(b|a)) \rangle \neq R$.

In this case there are more commands $y \in \mathbf{t}P\lceil \mathbf{c}P$ that lead to the same responses ($y = aa$ as well as $y = ab$ leads to the response de). To find a solution not all these commands are needed. Just one will do, but computing $F(P, L)$ leads to all possible commands. □

It is easily seen that if in P all responses can be found by applying a unique command only, all solutions of CODE will be of the form $F(P, L)$. A system P with this property is called observable:

Definition 6 P *is observable with respect to some response alphabet* E, *to be denoted by* $\mathbf{observable}_E(P)$, *if (with* $C = \mathbf{a}P \setminus E$)

$$(\forall x, y : x \in \mathbf{b}P \wedge y \in \mathbf{b}P : x\lceil E = y\lceil E \Rightarrow x\lceil C = y\lceil C)$$

Example 8 The following system $P = \langle \{a, b, d, e\}, (ae|ad|be) \rangle$ is not observable with respect to $\mathbf{e}P = \{d, e\}$: we have that $x = ae$ and $y = be$ both satisfy $x\lceil \mathbf{e}P = y\lceil \mathbf{e}P$ but $x\lceil \mathbf{c}P \neq y\lceil \mathbf{c}P$. However both $P' = \langle \{a, b, d, e\}, (ad|be) \rangle$ and $P'' = \langle \{a, b, d, e\}, (ae|ad) \rangle$ are observable with respect to the same exogenous alphabet. P'' has no solution for CODE with $L_{min} = L_{max} = \langle \{d, e\}, \{\epsilon, e\}, \{e\} \rangle$. P' has one effective solution, namely $R = \langle \{a, b\}, \{\epsilon, b\}, \{b\} \rangle$. □

Lemma 1 *For realistic systems* P *we have:*

$$\mathbf{observable}_{\mathbf{e}P}(P)$$

$$\Rightarrow$$

$$(\forall R : R \text{ is a solution of CODE} : F(P, P \rceil\lceil R) \cap P\lceil \mathbf{c}P = R \cap P\lceil \mathbf{c}P)$$

Proof: We only have to prove that $F(P, P \rceil\!\lceil R) \cap P\lceil cP \subseteq R \cap P\lceil cP$, because the other inclusion immediately follows from property 8 (b). We prove that the behaviour set of the first system is included in the second:

$x \in \mathbf{b}(F(P, P \rceil\!\lceil P) \cap P\lceil cP)$

$\Leftrightarrow \quad$ [definition of $\cap$]

$x \in \mathbf{b}P\lceil cP \wedge x \in \mathbf{b}(F(P, P \rceil\!\lceil R)$

$\Leftrightarrow \quad$ [definition of $F(P, L)$]

$x \in \mathbf{b}P\lceil cP \wedge x \in (cP)^* \setminus \mathbf{b}(P \rceil\!\lceil \sim(P \rceil\!\lceil R))$

$\Leftrightarrow \quad$ [$x \in \mathbf{b}P\lceil cP \Rightarrow x \in (cP)^*$]

$x \in \mathbf{b}P\lceil cP \wedge x \notin \mathbf{b}(P \rceil\!\lceil \sim(P \rceil\!\lceil R))$

$\Leftrightarrow \quad$ [definition of $\rceil\!\lceil$]

$x \in \mathbf{b}P\lceil cP \wedge (\forall y : y \in \mathbf{b}P \wedge y\lceil eP \in \mathbf{b}(\sim(P \rceil\!\lceil R)) : y\lceil cP \neq x)$

$\Leftrightarrow \quad$ [definition of $\sim$]

$x \in \mathbf{b}P\lceil cP \wedge (\forall y : y \in \mathbf{b}P \wedge y\lceil eP \in (eP)^* \setminus \mathbf{b}(P \rceil\!\lceil R) : y\lceil cP \neq x)$

$\Leftrightarrow \quad$ [$y \in \mathbf{b}P \Rightarrow y\lceil eP \in (eP)^*$]

$x \in \mathbf{b}P\lceil cP \wedge (\forall y : y \in \mathbf{b}P \wedge y\lceil eP \notin \mathbf{b}(P \rceil\!\lceil R) : y\lceil cP \neq x)$

$\Leftrightarrow$

$x \in \mathbf{b}P\lceil cP \wedge (\forall y : y \in \mathbf{b}P \wedge y\lceil cP = x : y\lceil eP \in \mathbf{b}(P \rceil\!\lceil R))$

$\Leftrightarrow \quad$ [definition of $\rceil\!\lceil$]

$x \in \mathbf{b}P\lceil cP \wedge (\forall y: y \in \mathbf{b}P \wedge y\lceil cP = x$
$\qquad : (\exists z : z \in \mathbf{b}P \wedge z\lceil cP \in \mathbf{b}R : z\lceil eP = y\lceil eP))$

$\Leftrightarrow$

$x \in \mathbf{b}P\lceil cP \wedge (\forall y: y \in \mathbf{b}P \wedge y\lceil cP = x$
$\qquad : (\exists z : z\lceil eP = y\lceil eP \wedge z \in \mathbf{b}P : z\lceil cP \in \mathbf{b}R))$

$\Rightarrow \quad$ [$z\lceil eP = y\lceil eP \Rightarrow z\lceil cP = y\lceil cP$, P is observable]

$x \in \mathbf{b}P\lceil cP \wedge (\forall y : y \in \mathbf{b}P \wedge y\lceil cP = x : y\lceil cP \in \mathbf{b}R)$

$\Rightarrow$

$x \in \mathbf{b}P\lceil cP \wedge x \in \mathbf{b}R$

$\Leftrightarrow$

$x \in \mathbf{b}(R \cap P\lceil cP)$

The same can be proven for the task set, because for realistic systems the observability on the behaviours also holds for all tasks.

6.1 CODE FOR OBSERVABLE SYSTEMS

From lemma 1 it is clear that every effective part of a solution of CODE for an observable discrete system P is equal to the effective part of a solution of the form $F(P, L)$.

From property 8 (a) we see that

$$L_{min} \subseteq L \Rightarrow F(P, L_{min}) \subseteq F(P, L)$$

holds for every L satisfying the minmax condition. If P is observable, all effective parts of the solutions are of the form $F(P, L) \cap P\lceil cP$. We conclude that for an observable P the effective part of the solution $F(P, L_{min})$ is minimal:

Theorem 3 *If CODE has a solution, then $F(P, L_{max}) \cap P\lceil cP$ is the largest solution contained in $P\lceil cP$. If in addition P is realistic and observable, then $F(P, L_{min}) \cap P\lceil cP$ is the smallest solution contained in $P\lceil cP$.*

Proof: It is clear that $F(P, L_{max}) \cap P\lceil \mathbf{c}P$ is the largest solution that is contained in $P\lceil \mathbf{c}P$. Suppose R is some solution of CODE and $R \subseteq P\lceil \mathbf{c}P$, then we have

$$F(P, L_{min}) \cap P\lceil \mathbf{c}P$$
$$\subseteq \quad [\ R \text{ is a solution, so } L_{min} \subseteq P\rceil\lceil R\ ]$$
$$F(P, P\rceil\lceil R) \cap P\lceil \mathbf{c}P$$
$$= \quad [\ P \text{ is realistic and observable}\ ]$$
$$R \cap P\lceil \mathbf{c}P$$
$$= \quad [\ R \subseteq P\lceil \mathbf{c}P\ ]$$
$$R$$

so $F(P, L_{min}) \cap P\lceil \mathbf{c}P$ is the smallest solution

7. LOCKED SYSTEMS

Because we have defined a DES using two sets of sequences of events, the behaviour and the task set, we are able to tell if at some point in behaviour a system is able to perform a completed task. If at some point no task can be completed any more, we say the system is *locked*. This (informal) definition of lock contains the well-known deadlock: not being able to do a next event (and not having completed a task), but also, what we will call, livelock: being able to do a next event at all times, but never being able to perform a task any more. Each system that contains deadlock cases, livelock cases, or combinations is called a system with the possibility of lock. The term deadlock appears in DES-literature more often as blocking.

Definition 7 *We say a DES has the possibility to lock if the set*

$$\mathbf{lock}(P) = \mathbf{b}P \setminus \mathbf{pref}(\mathbf{t}P)$$

is not empty.

Notice that $\mathbf{lock}(P)$ contains those behaviour that cannot be completed to become a task of P. Similar we say a connection of systems P and R may lock if $\mathbf{lock}(P \parallel R) \neq \emptyset$.

Example 9 Consider $P = \langle \{a, b\}, \{\epsilon, a, aa, ab, abb, abbb, abbbb, \ldots\}, \{ab\}\rangle$. Then we have $\mathbf{lock}(P) = \{aa, abb, abbb, \ldots\}$. Behaviour aa is deadlock: P cannot do anything more and has not completed a task after aa. Behaviour $abbb\ldots$ is livelock: although system P can do event b over and over again, it will never complete a task. □

The controller $\mathbf{des}(F(P, L_{max}))$ may lead to a connection with P that can lock. Therefore, we introduce a second control problem, similar to the previous one, but with an extra demand: the connection $P \parallel R$ of plant P and controller R should be free of lock. However, in order to find a solution for this problem, we first introduce some other problems:

Control problem 2 (LFC) *Given two systems P and R with $\mathbf{lock}(P \parallel R) \neq \emptyset$, find the greatest subsystem of R that has a lock free connection with P.*

In order to solve this second problem we recall some results from [2, 9, 10, 11]. Given some realistic systems P and R with $\mathbf{lock}(P \parallel R) \neq \emptyset$, we can construct a subsystem $R_{lf} \subseteq R$ such that $\mathbf{lock}(P \parallel R_{lf}) = \emptyset$. The following operator leads to this subsystem:

$$L(P, R) = R \setminus \mathbf{lock}(P \parallel R) \lceil \mathbf{a}R$$

where $R \setminus T$ is defined by

$$R \setminus T = \mathbf{des}(\langle \mathbf{a}R, \mathbf{b}R \setminus T, \mathbf{t}R \setminus T \rangle)$$

Example 10 Consider $P = \langle \{a, b\}, (a(ba|ab)) \rangle$ and $T = \{aa\}$. Then we have

$$
\begin{aligned}
P &= \langle \{a, b, \}, \{\epsilon, a, ab, aa, aba, aab\}, \{aba, aab\} \rangle \\
P \setminus T &= \mathbf{des}(\langle \{a, b\}, \{\epsilon, a, ab, aba, aab\}, \{aba, aab\} \rangle) \\
&= \langle \{a, b\}, \{\epsilon, a, ab, aba\}, \{aba\} \rangle \\
&= \langle \{a, b\}, (aba) \rangle
\end{aligned}
$$

$\square$

Although T is only some trace set, we will denote the system $\langle \mathbf{a}P, T, T \rangle$ also by T. We have:

Property 9

$$P \setminus T = P \cap \sim T = P \parallel \sim T$$

Property 10
(a) $L(P, R) \subseteq R$
(b) $L(P, S) = S \Leftrightarrow \mathbf{lock}(P \parallel S) = \emptyset$
(c) $L(P, S) = S \wedge S \subseteq R \Rightarrow S \subseteq L(P, R)$

The greatest subsystem of R that leads to a lock free connection with P now is the greatest fix-point of $L(P, R)$, i.e., we iteratively compute:

$$
\begin{aligned}
R_0 &= R \\
R_{i+1} &= L(P, R_i)
\end{aligned}
$$

until that fix-point is reached. Let $\Lambda(P, R)$ denote that fix-point, then we have:

Property 11
(a) $\mathbf{lock}(P \parallel \Lambda(P, R)) = \emptyset$
(b) $S = L(P, S) \wedge S \subseteq R \Rightarrow S \subseteq \Lambda(P, R)$

So we have:

Theorem 4 $\Lambda(P, R)$ *is the greatest subsystem of R for which the connection with P is free of lock.*

(For the proofs of the above properties and theorem we refer to [2] and [11].)

The above results can be used if R is known. Sometimes we only have P and want to control P using $\mathbf{c}P$ in such a way that no lock can occur. This leads to the following additional problem:

Control problem 3 (LF) *Given some system P with $\mathbf{lock}(P) \neq \emptyset$ and $\mathbf{c}P \subseteq \mathbf{a}P$ the set of control-events of P, find a controller R, with $\mathbf{a}R = \mathbf{c}P$, such that the connection $P \parallel R$ is free of lock and R is minimal restrictive.*

The controller R must be minimal restrictive, i.e., should be as general as possible. This suggests that we start with the most general controller that is possible: $R = \mathbf{chaos}(\mathbf{c}P)$. Now applying the algorithm for solving the LFC-problem, starting with $R_0 = R$ we find a fix-point that turns out to be the greatest subsystem of R that leads to a lock free connection with P and, therefore, is minimal restrictive:

Theorem 5 *If $\Lambda(P, \mathbf{chaos}(\mathbf{c}P)) = \mathbf{empty}(\mathbf{c}P)$ no minimal restrictive controller for the LF-problem can be found, otherwise $\Lambda(P, \mathbf{chaos}(\mathbf{c}P))$ is the solution.*

We refer to [10] for a more thorough discussion on this problem.

8. LOCK FREE CONTROL PROBLEM

The solutions of the previous control problems can now easily be combined to give a solution for the following control problem:

Control problem 4 (LFCODE) *Given a system P, control events $\mathbf{c}P \subseteq \mathbf{a}P$ and two systems $L_{min} \subseteq L_{max} \subseteq \mathbf{chaos}(\mathbf{e}P)$ (with $\mathbf{e}P = \mathbf{a}P \setminus \mathbf{c}P$), find a controller R with $\mathbf{a}R = \mathbf{c}P$ such that $L_{min} \subseteq P \rceil\!\lceil R \subseteq L_{max}$ and $\mathbf{lock}(P \parallel R) = \emptyset$.*

This solution can now easily be found by combining the solutions of CODE and LFC: first we compute $R = \mathbf{des}(F(P, L_{max}))$ to find the greatest controller such that the connection $P \rceil\!\lceil R$ satisfies the minmax condition. Next, we compute the greatest subsystem of this R that also leads to a lock free connection with P:

Theorem 6 *A lock free controller for the control problem can be found if and only if*

$$L_{min} \subseteq P \rceil\!\lceil \Lambda(P, \mathbf{des}(F(P, L_{max})))$$

The greatest solution then is given by $\Lambda(P, \mathbf{des}(F(P, L_{max})))$

9. EFFECTIVELY COMPUTABLE

State graphs can be used to get an automaton-like representation of DESs. It turns out that all operators used to compute solutions for our control problems can easily be translated into operators on these automatons, which enables us to compute the controllers effectively if the original systems we start with are regular (i.e., contain a finite number of states in there automaton representation). In [12] an example of the use of this theory can be found.

10. CONCLUSION

In this paper we have presented an elegant way of modelling discrete event systems, enabling us to specify and solve control problems. The controller can easily be computed using the reflection operator which, temporally, brings us beyond the scope of a realistic discrete event system.

REFERENCES

[1] J.L.A. van de Snepscheut. *Trace theory and VLSI design.* Lecture notes in computer science, nr. 200. Springer Verlag, 1985.

[2] R. Smedinga. Locked discrete event systems. Technical Report CS9002, Department of computing science, University of Groningen, 1990.

[3] R. Smedinga. The reflection operator in discrete event systems. Technical Report CS9201, Department of computing science, University of Groningen, 1992.

[4] R. Smedinga. Discrete event systems. course-notes, Department of computing science, University of Groningen, 1992.

[5] S. Balemi. *Control of discrete event systems: theory and application.* PhD thesis, Swiss federal institute of technology, Zürich, May 1992.

[6] P.J. Ramadge and W.M. Wonham. Supervisory control of a class of discrete event processes. *SIAM journal on Control and optimalisation,* 25 (1), 1987. See also: systems control group report 8515, department of electrical engineering, University of Toronto.

[7] P.J. Ramadge and W.M. Wonham. The control of discrete event systems. *Proceedings of the IEEE,* 77(1), 1989.

[8] T.Verhoeff. Factorization in process domains. Technical Report 91/03, department of mathematics and computing science, Eindhoven university of technology, 1991.

[9] R. Smedinga. Discrete event systems: deadlock, livelock, and livedeadlock. In U. Jaaksoo and V.I. Utkin, editors, *Automatic Control, world congress 1990, 13–17 August, proceedings of 11th IFAC world congress,* volume III, Tallinn, Estonia, USSR, 1991. Pergamon Press.

[10] R. Smedinga. An effective way to undo a discrete event system of its (dead)lock. In *Preprints of proceedings of the IFAC Symposium on Design Methods for Control Systems,* Zürich, Switzerland, 4–6 Sep. 1991.

[11] R. Smedinga. Locked discrete event systems: how to model and how to unlock. submitted to the journal of Discrete Event Dynamical Systems, 1992.

[12] R. Smedinga. The workshop exercise using a trace theory based setting. *In this volume,* pages 167–172.

A Minimally Restrictive Policy for Deadlock Avoidance in a Class of FMS*

Yitzhak Brave[†] Dominique Bonvin[‡]

Abstract. In recent years, the scheduling and control of Flexible Manufacturing Systems (FMS) has received considerable attention. Due to complexity considerations, heuristic solutions are often the norm for real-time scheduling and implementation of FMS control activities. However, heuristic solutions often lack consideration of overall system implications, and various practical difficulties may arise in the control of unmanned FMS.

The purpose of this research is to address deadlock situations that may arise in unmanned FMS. Petri nets are used to model the sequencing of material through the workstations and the limited capacity of resources available in the FMS. The model is then used for defining restriction policies (i.e., policies that disallow some of the options of the real-time resource allocation strategy) for deadlock avoidance. We propose an on-line restriction policy that avoids deadlock situations, while enabling completion of all active jobs in the FMS. The proposed algorithm is minimally restrictive and suitable for real-time implementation.

1. INTRODUCTION

Flexible Manufacturing Systems (FMS) are characterized by the availability of multiple workstations at which generic operations are performed and of a programmable transportation system connecting the workstations. Using the transportation system, material is routed through the workstations to manufacture various products. Each product has a predefined operation sequence that determines the order in which resources must be assigned to the manufacturing process of the product. Due to complexity considerations, heuristic real-time resource allocation policies are often the norm in FMS. Heuristic solutions, however, often lack considerations of overall system implications, and various practical difficulties may arise in the control of (unmanned) FMS. One specific problem which is addresses in this paper is deadlock situations.

Deadlock occurs when two or more jobs (production processes) are holding a number of resources, and no process can proceed due to the fact that each is waiting for resources to be released by another. There are four necessary conditions that must hold if deadlock is to occur [1]:

1. *mutual exclusion* – only one process can use a resource at once;

2. *hold and wait* – processes hold onto resources while waiting for others;

---

*This work was supported by the fonds national suisse de la recherche scientifique

[†]Institut d'Automatique, Ecole Polytechnique Fédérale de Lausanne (EPFL), CH-1015, Lausanne, Suisse, brave@elia.epfl.ch

[‡]Institut d'Automatique, Ecole Polytechnique Fédérale de Lausanne (EPFL), CH-1015, Lausanne, Suisse, bonvin@elia.epfl.ch

3. *no preemption* – a resource can only be released voluntarily by a process, and;

4. *circular wait* – a circular chain of processes in which each process holds resources which are requested by the next process in the chain.

If an FMS is to be reliable it must address the issue of deadlock. There are three possible approaches:

1. *Deadlock prevention* – at least one of the four conditions required for deadlock never occurs in the system.

2. *Deadlock avoidance* – an algorithm can be constructed which allows all four conditions necessary for deadlock to occur but which also ensures that the system never enters a deadlock state.

3. *Deadlock detection and recovery* – problems of deadlock are ignored until the system enters a deadlock state, at which some corrective action is taken.

In the FMS application considered in the present paper, namely an FMS with a collection of workstations and local buffers, in which material can be removed from any buffer only when one of the other buffers is not full, the first three conditions necessary for deadlock to occur are always present. Taking this fact into account, as well as our intention to design reliable unmanned FMS, the second approach for dealing with deadlock situations is preferred. Although it is easy to avoid circular chain of processes waiting for resources (for instance, by merely allowing only one process in the system at a time), a good deadlock avoidance algorithm (DAA) should not affect performance measure such as resource utilization and throughput. DAAs for computer operating systems (e.g., banker's algorithm [2]) are not efficient for FMS applications because they do not make use of the information about the order in which resources will be requested or released ([3, 4]). Recently, a much less restrictive DAA has been proposed by Banaszak and Krogh [3]. In their approach, each production sequence is partitioned into *zones* of shared and unshared resources. The advancement of jobs in the system is then monitored by rules that limit the number of busy resources in each zone, so as to guarantee that deadlock will never be reached. These rules, however, do not provide the set of minimal restrictions that can be placed on resource allocation decisions for ensuring deadlock avoidance.

In this paper, we study restriction policies (i.e., policies that disallow some of the options of the real-time resource allocation strategy) for deadlock avoidance. Petri nets (PN) - a formalism that has been used extensively to describe, analyze and implement the control of FMS [5] - are used to model the sequencing of material through the workstations and the limited capacity of resources available in the FMS. The model is then used for defining an on-line restriction policy that avoids deadlock situations, while enabling completion of all active jobs in the FMS. The proposed algorithm is minimally restrictive and suitable for real-time implementation.The Petri net model of the FMS is described and illustrated in section 2.. In section 3., we formally define deadlock situations, introduce resource dependency graphs and state our DAA. The correctness proof of this algorithm is presented in section 4., and its relation to existing DAAs is discussed in section 5.. An extended version of this report can be found in [6].

2. PETRI NET MODEL

To model the flow of jobs in an FMS we shall use a PN model similar to that in [3]. This model reflects only the aspects that are relevant to the deadlock avoidance problem. We consider a FMS consisting of a set of resources R, where each resource $r \in R$ has capacity C_r (a positive integer greater than one). The set of products to be produced by the FMS is denoted by Q. We assume that for each product $q \in Q$, the operation sequence to produce q defines a sequence of resource requirements $\mathbf{r_q} = \{r_q(1), \cdots, r_q(L_q)\}$, where L_q is the number of steps in the operation sequence of product q, and $r_q(j) \in R$ for $j = 1, \cdots, L_q$. It is assumed that each operation requires at most one unit of one resource, and that a unit of resource can be involved in one operation at a time. To represent the progress of jobs in the FMS, a sequence of places $\mathbf{p_q} = \{p_q(0), \cdots, p_q(L_q + 1)\}$, called *production sequence*, is associated with each sequence of resource requirements $\mathbf{r_q}$. Tokens in place $p_q(0)$ represent orders for product q waiting to be initiated, and tokens in place $p_q(L_q + 1)$ represent completed orders for q. Tokens in an intermediate place $p_q(j)$ represent unfinished orders for q being at their j-th step of production, where each of these unfinished orders holds one resource of type $r_q(j)$. No resources are required by jobs represented by tokens in places $p_q(0)$ and $p_q(L_q + 1)$.

The advancement of jobs through the FMS is modeled by the sequencing of tokens through the *production places* $p_q(j)$, as illustrated in figure 1.

Firing transition $t_q(0)$ corresponds to the arrival of a new order to execute a job for product q, whereas firing subsequent transition $t_q(j)$ corresponds to the progress of a job for product q from step j to step $j + 1$ in the production sequence of q. For each resource r we assign a *resource place* named r. Tokens in place r indicate available resources of type r. If resource r is required at the j-th production step of product q, i.e. $r_q(j) = r$, there is an input arc from place r to transition $t_q(j)$, and an output arc from $t_q(j+1)$ to r. Thus, if $t_q(j+1)$ fires, one resource of type $r_q(j)$ is released and one resource of type $r_q(j + 1)$ is occupied. We shall use the notation $(p, r) \xrightarrow{t} (p', r')$ for representing a transition t that has a pair of input places (p, r) and a pair of output places (p', r'), where each pair consists of a production place and a resource place.

Consider the FMS illustrated in figure 2 in which three workstations are serviced by a robotic device. The single robot is used to load/unload parts and to move parts for processing between the various workstations (W_1, W_2, W_3). Each workstation W_j has an input buffer I_j to hold workpieces to be processed on machine M_j and an output buffer O_j to hold workpieces for which the job step has been completed. Each buffer is assumed to have three available slots. Jobs enter the system at random and require multiple machines for completion in a specific operation sequence. For example, suppose a product of type 1 is manufactured by sequencing material through workstations W_3, W_1, W_2 in order. For this product, the resource requirement sequence is given by $\mathbf{r_1} = \{I_3, O_3, I_1, O_1, I_2, O_2\}$ (for

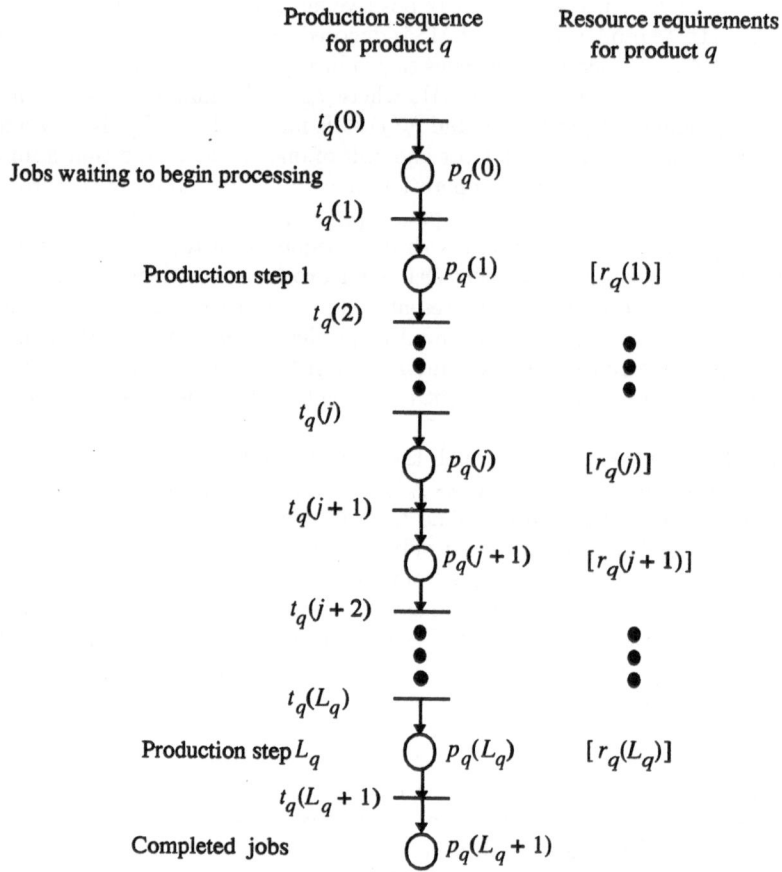

Figure 1: Petri net representation for production sequence of product q.

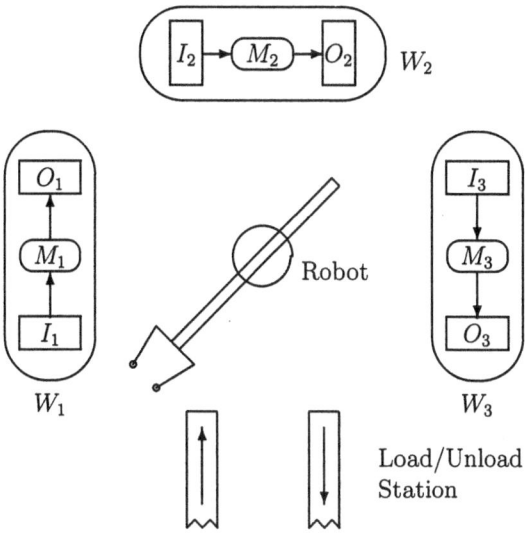

Figure 2: FMS with three workstations.

the purpose of deadlock analysis there is no need to include the single robot and the single machine in each cell in the resource requirement sequence). Assume further that the FMS produces a product of type 2 by moving material through W_1, W_3, W_1. The PN model of these production sequences is shown in figure 3, in which, for sake of clarity, transition labels are used to identify the resource transitions with the production transitions. Thus, for instance, the input and output resource places of transition $t_1(6)$ (see production sequence of product 1) are O_2 and I_2, respectively.

In general, the PN model of the process flow in an FMS is represented by a fivetuple (P, T, I, O, M_0), where P is a finite set of places, T is a finite set of transitions, $I \subseteq P \times T$ is a set of transition input arcs, $O \subseteq T \times P$ is a set of transition output arcs, and $M_0 : P \to N$ is the initial marking of the PN, where N is the set of nonnegative integers. For the model described above we have:

$$
\begin{aligned}
P &= R \cup \{p_q(j) \mid q \in Q,\ j = 0, \dots, L_q + 1\}, \\
T &= \{t_q(j) \mid q \in Q,\ j = 0, \dots, L_q + 1\}, \\
I &= \{(p_q(j-1), t_q(j)), (r_q(j), t_q(j)) \mid q \in Q,\ j = 1, \dots, L_q + 1\}, \\
O &= \{(t_q(j), p_q(j)) \mid q \in Q,\ j = 1, \dots, L_q + 1\} \cup \\
&\qquad \{(t_q(j+1), r_q(j)) \mid q \in Q,\ j = 1, \dots, L_q\}, \\
M_0(p) &= \begin{cases} C_p & \text{if } p \in R \\ 0 & \text{otherwise,} \end{cases}
\end{aligned}
$$

where for $p \in R$, C_p is the total number of resource p available in the system. The initial marking M_0 represents a system with no job in progress. The reader

Production sequence of product type 1	Resources	Production sequence of product type 2

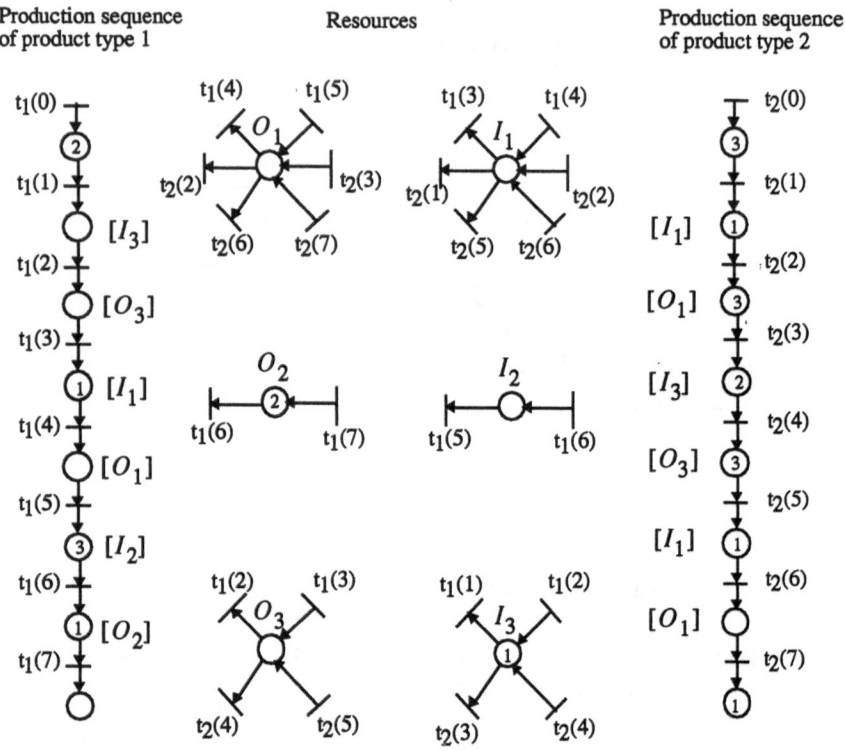

Figure 3: PN model of resource requirement sequences for product types 1 and 2 for the FMS in figure 2.

is referred to [7] for the standard definition of transition firing rules. The set of all markings reachable from the initial marking M_0 is denoted $\mathcal{R}(M_0)$.

In figure 3, the current marking M represents a situation with two orders for product type 1, three orders for product type 2, one completed product of type 2, and five products of type 1 and ten products of type 2 being at various intermediate stages of production. Also, resources I_3 and O_2 have one and two free slots, respectively, whereas the rest of the resources are full or exhausted (recall that, in this example, $C_r = 3$ for every $r \in R$). Under marking M, only transitions $t_1(1)$, $t_2(3)$ and $t_1(6)$ can fire (we have neglected here transitions, such as $t_1(0)$ and $t_1(7)$, whose firing do not depend on the availability of free resources). The first two transitions, namely $t_1(1)$ and $t_2(3)$, compete for the same resource - I_3 - which has only one free slot. Now, if transition $t_1(1)$ fires, buffer I_3 becomes full, thereby causing transitions $t_2(j)$, $1 \leq j \leq 6$, as well as $t_1(1)$, $t_1(2)$ and $t_1(4)$ to enter a deadlock situation. It can easily be verified that, once the unfinished jobs in buffers I_2 and O_2 are completed, no other jobs can be either processed or accepted.

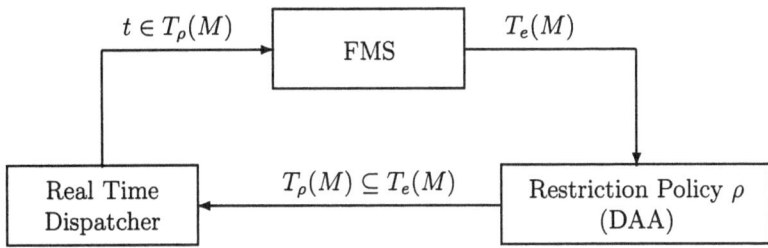

Figure 4: Interrelation between the FMS, the real-time dispatcher and the restriction policy.

The recovery from deadlock situation might be expensive, particularly during unmanned night operation of an FMS. Consequently, one may be interested in finding restriction policies that prevent such situations, while enabling the completion of any unfinished job in the FMS. The interrelation between the FMS, the real-time dispatcher and the restriction policy is assumed to be as follows (see figure 4 and definition 1 below). The restriction policy module, denoted ρ, reads the current marking M of the PN (representing the current state of the FMS) and specifies a subset of transitions that are enabled under M. The real-time dispatcher then fires one of the transitions specified by ρ, thereby causing the PN to enter a new marking. The updated marking is read by ρ and a new decision cycle is initiated. This procedure is repeated as long as there are new or unfinished jobs in the FMS. To facilitate the discussion on deadlock, the following notation are introduced.

Definition 1 *Given a marking M and a transition $(p,r) \overset{t}{\twoheadmapsto} (p',r')$,*

- *transition t is* production enabled *if $M(p) \geq 1$ (means that a job is currently in the production step preceding the transition t);*

- *transition t is* resource enabled *under M if $M(r) \geq 1$ (means that the resource required in the next production step is available);*

- *transition t is* enabled *under M if it is production enabled and resource enabled ;*

- *$T_e(M)$ denotes the set of enabled transitions under M;*

- *$T_\rho(M)$ denotes a subset of enabled transitions that may fire at marking M under a restriction policy ρ;*

- *$\mathcal{R}_\rho(M)$ denotes the set of all markings reachable from the marking M under a restriction policy ρ;*

3.   PROPOSED ALGORITHM

Our objective is to design a restriction policy ρ which guarantees that at every reachable marking, any production enabled transition can eventually be fired.

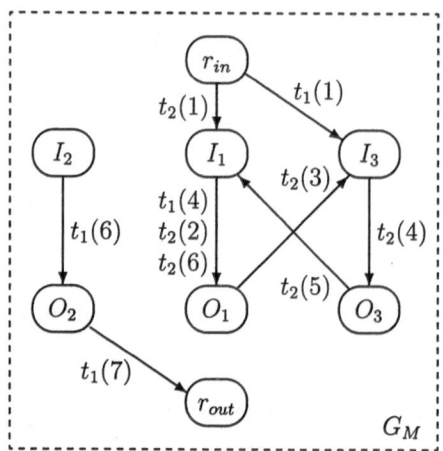

Figure 5: Resource dependency graph of the PN model depicted in figure 3.

Formally, we have the following:

Definition 2 *Given sets of resources R and products Q, a PN for the production sequences, a restriction policy ρ, and a marking $M \in \mathcal{R}_\rho(M_0)$, a subset of production enabled transitions $T' \subseteq T$ is said to be in ρ-restricted deadlock under M if $T' \cap T_\rho(M') = \emptyset$ for every $M' \in \mathcal{R}_\rho(M)$. If the latter condition holds with respect to any restriction policy, the term $T\rho$-restrictedU will be omitted.*

For defining our DAA, we first introduce the resource dependency graph that represents the resources required at the next step of each active job in an FMS. Recall that in section 2. it was assumed that no resources are required by orders for products and completed orders represented by tokens in places $p_q(0)$ and $p_q(L_q + 1)$, respectively (see also figure 1). Alternatively, we may assume that these Tend-productsU require infinite-capacity dummy resources r_{in} and r_{out}, respectively. Accordingly, the set R of all (real) resources is extended to include these dummy resources.

Definition 3 *Given sets of resources R and products Q, a PN for the production sequences, and marking M, define the resource dependency graph $G_M = (R, E_M)$ with node set R and edge set E_M, where an edge $r' \xrightarrow{t} r$ is an element of E_M iff there exists a production enabled transition $(p, r) \xrightarrow{t} (p', r')$.*

For instance, figure 5 shows the resource dependency graph of the PN model depicted in figure 3. Notice that there are three edges going from node I_1 to O_1.

In the remaining of this paper we shall assume that the capacity of every resource (buffer) is greater than one (an assumption which is not a restriction in many applications). We are ready now to pose our restriction policy for deadlock avoidance.

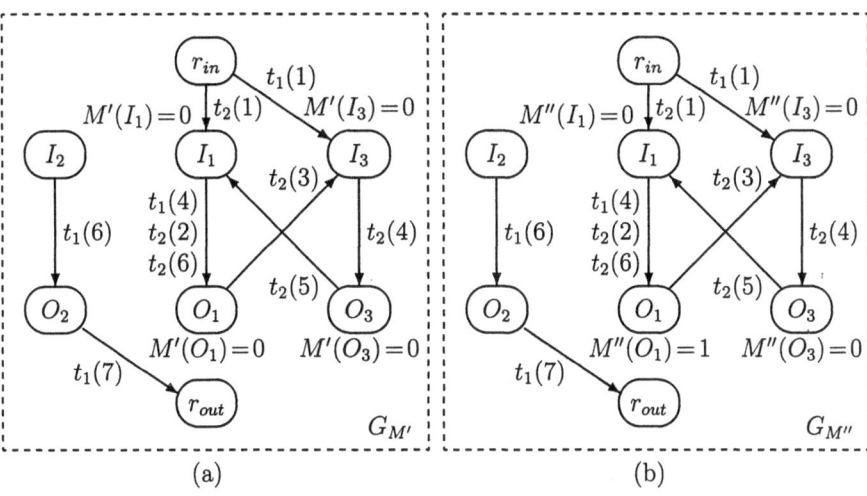

Figure 6: Resource dependency graph of the PN model depicted in figure 3 after the firing of transition (a) $t_1(1)$. (b) $t_2(3)$.

Deadlock avoidance algorithm α: *Let M the current marking of the PN. For every enabled transition t, with associated edge $r' \xrightarrow{t} r$, perform the following steps:*

1. *Compute the marking M' which is obtained by the exclusive firing of t at M and construct its associated resource dependency graph $G_{M'}$.*

2. *Compute the set $\mathrm{Reach}(G_{M'}, r)$ of all nodes in $G_{M'}$ that are reachable from node r.*

3. *Disallow transition t iff every node (resource) in this reach is fully busy under marking M', that is, iff the following disabling condition holds*

$$\forall s \in \mathrm{Reach}(G_{M'}, r), \qquad M'(s) = 0. \tag{1}$$

■

We shall demonstrate DAA α on the FMS in figure 2. Suppose the current marking is given by the token distribution in figure 3, and consider the transition $t_1(1)$. First, construct the resource dependency graph $G_{M'}$ (see figure 1a), where M' is the marking that is reached after the firing of $t_1(1)$ at M. Since every resource in the set $\mathrm{Reach}(G_{M'}, I_3) = \{I_3, O_3, I_1, O_1\}$ is fully busy, we conclude that DAA α disallows $t_1(1)$ under marking M. Remark that the latter conclusion coincides with the observation above definition 1, regarding the consequences of firing transition $t_1(1)$. If, however, this procedure is repeated with respect to transition $t_2(3)$ instead of $t_1(1)$ (see figure 6b), then the fact that resource O_1 has one free slot and it is reachable from I_3 would imply the enabling of $t_2(3)$ under DAA α.

DAA α is an on-line restriction policy. That is, the test whether a marking in the reachability graph of the PN may lead to a deadlock situation is performed only if this marking is actually reached. This way, a computationally expensive a priori search of the reachability graph can be avoided. Whenever there is a change in the marking of the PN, the DAA is re-executed. The decision as to whether a given enabled transition should be disabled or allowed does not depend on the decision regarding other transitions. Thus, the sequential steps 1-3 can be carried out simultaneously with regards to all enabled transitions, thereby reducing the disabling period of transitions that does not satisfy condition (1).

The resource dependency graph is updated during execution as follows. Assume that the firing of transition $(p, r) \xrightarrow{t} \mapsto (p', r')$ at marking M causes the PN to enter marking M', and let $r' \xrightarrow{t} r$ be the edge (in resource dependency graph G_M) that is associated with t. Resource dependency graph $G_{M'}$ may differ from G_M in no more than two edges. Specifically, the edge $r' \xrightarrow{t} r$ may be deleted in $G_{M'}$, and a new edge starting at node r may be added to $G_{M'}$. Thus,

$$G_M - \{r' \xrightarrow{t} r\} \subseteq G_{M'} \subseteq G_M \cup \{r \xrightarrow{t'} r''\} \tag{2}$$

Furthermore,

$$M'(r') = M(r') + 1, \qquad M'(r) = M(r) - 1, \tag{3}$$

and $M'(s) = M(s)$ for any other node $s \in R$.

The complexity of DAA α is $O(|T|^2)$, where $|T|$ is the size of the transition set in the PN representation. At initialization, the resource dependency graph has no edges. Then, after every marking transition, the resource dependency graph is updated with complexity $O(1)$. This is also the complexity of step 1 in DAA α. Thus, the overall complexity of this algorithm follows from the fact that the reachability computation of complexity $O(|T|)$ in step 2 is performed at most $|T|$ times.

4. CORRECTNESS PROOF OF DAA α

Disabling condition (1) plays a key role in the proposed restriction policy. In fact, the next lemma shows that condition (1) is universal in the sense that *any* restriction policy for deadlock avoidance must guarantee that this condition is never satisfied.

Lemma 1 *If for some marking M' and some node $r \in R$ condition (1) is satisfied then every production enabled transition whose input resource place belongs to* Reach$(G_{M'}, r)$ *is in deadlock.*

Proof: Suppose condition (1) holds for a marking M' and a node $r \in R$, and consider an edge of the form $s \xrightarrow{t} s'$ with $s' \in$ Reach$(G_{M'}, r)$. Transition t can become enabled iff a transition that can release resource s' fires, for instance, a transition t' with associated edge $s' \xrightarrow{t'} s''$. This, however, is impossible since $s'' \in$ Reach$(G_{M'}, r)$ and, therefore, by (1), $M(s'') = 0$. Thus, we conclude that the resources in Reach$(G_{M'}, r)$ will remain indefinitely fully busy for any marking $M \in \mathcal{R}(M')$. ∎

The next lemma states that the situation described by condition (1) indeed cannot occur under DAA α (due to space limitations, the proofs of the following lemmas are omitted).

Lemma 2 *There is no $M' \in \mathcal{R}_\alpha(M_0)$ and $r \in R$ such that condition (1) holds.*

So far, it has been shown that, under DAA α, there is always a resource which has at least one free slot; this is, in fact, an immediate consequence of Lemma 2. However, the latter conclusion says nothing about the possibility of using a specific resource. In the sequel we shall show that even resources with only one free slot can always be used by firing internal transitions, a specific type of transitions which is defined as follows.

Definition 4 *A production place is called* internal *if it is neither the first nor the last place in its production sequence. A transition is called* internal *if its input production place is internal.*

Thus, a token in an internal production place represents an order for product that the FMS has started producing, but is not completed yet. For instance, in figure 1, only places $p_q(0)$ and $p_q(L_q + 1)$, and transitions $t_q(0)$ and $t_q(1)$ are not internal.

Assume now that some resource has only one free slot and that there is a subset $\hat{T}$ of transitions that compete for this slot. The next lemma states that DAA α never disallows all internal transitions in $\hat{T}$.

Lemma 3 *Let $M \in \mathcal{R}_\alpha(M_0)$ and let $t \in T_e(M) - T_\alpha(M)$. Then $T_\alpha(M)$ includes an internal transition whose input resource place is identical to that of t.*

The results of Lemmas 2 and 3 are summarized in the following lemma which claims that as long as there are unfinished jobs at the system, at least one of those jobs is allowed to be advanced under DAA α.

Lemma 4 *Let $M \in \mathcal{R}_\alpha(M_0)$ be a marking such that $M(p) \geq 1$ for some internal place p. Then DAA α allows at least one internal transition under M.*

Based on the aforementioned lemmas, we are able now to prove our first main result, which states that a deadlock situation can never occur under restriction policy α.

Theorem 1 *For any marking $M \in \mathcal{R}_\alpha(M_0)$, the set of transitions in α-restricted deadlock is empty.*

Proof: Assume, towards a contradiction, that for some marking $M \in \mathcal{R}_\alpha(M_0)$ there is a production enabled transition t which is in α-restricted deadlock under M. That is, for every marking $M' \in \mathcal{R}_\alpha(M_0)$ the transition t is either resource-disabled or disallowed under α. Assume further that from marking M, the system continues firing only internal transitions allowed under α. Notice that by the hypothesis that transition $(p, r) \xrightarrow{t} \!\!\!\!\!\! \not\;\;\;\; \mapsto (p', r')$ is in α-restricted deadlock under M it follows that the resource r, whose capacity is at least two slots by assumption, has at most one free slot (otherwise, disablement condition (1) does not hold). This, in turn, implies that r is occupied by an unfinished job, a fact represented by a token in an internal production place. By Lemma 4, it follows that the system can execute consecutively and indefinitely internal transitions. This, however, violates the fact that the number of tokens in internal places (which represent unfinished jobs) is bounded by the finite capacity of (real) resources, and is monotonically decreasing if only internal transitions are fired. In other words, a consecutive firing of internal transitions must eventually release all resources, thereby enabling every production enabled transition. The latter conclusion violates the contradiction assumption and completes the proof.   ∎

5.   RESTRICTIVENESS

In the previous section we have shown that DAA α is a restriction policy that avoids deadlock. Clearly, this policy is not unique. One can, for instance, obey a trivial policy that allows no more than one job at a time in the FMS. The latter policy, however, is very conservative in terms of resource utilization. In this section, we shall compare the restrictiveness of our policy and other on-line policies. In fact, we shall show that any other restriction policy that avoids deadlock is at least as restrictive as the proposed policy. To this end, we introduce the following restrictiveness measure.

Definition 5 *Let ρ and ρ' be two restriction policies that avoids deadlock. Policy ρ is said to be* less restrictive *than ρ' if*

$$\mathcal{R}_{\rho'}(M_0) \subseteq \mathcal{R}_{\rho}(M_0) . \tag{4}$$

If (4) holds for any restriction policy ρ' that avoids deadlock then ρ is called minimally restrictive.

Hence, if a policy ρ is less restrictive than ρ',

$$\forall M \in \mathcal{R}_{\rho'}(M_0), \qquad T_{\rho'}(M_0) \subseteq T_{\rho}(M_0) . \tag{5}$$

Practically, condition (5) means that, compared to policy ρ', policy ρ provides the real-time scheduler of the FMS more options and flexibility for its decision making process. This, in general, can increase productivity and improve resource utilization.

Let us now examine the restrictiveness of the proposed DAA α. Assume that for some marking M an enabled transition $(p, r) \overset{t}{\nrightarrow} (p', r')$ is disallowed under α. It follows from step 3 of DAA α that condition (1) is satisfied with respect to resource r, and therefore, by Lemma 1, transition t must be disallowed under *any* restriction policy that avoids deadlock. Thus, we have proved our second main result:

Theorem 2 *DAA α is minimally restrictive.*

The practical importance of this theorem follows from the fact that, in general, the less the restrictiveness, the more the resource utilization. We remark that, to the best of our knowledge, the proposed policy is strictly less restrictive than other on-line restriction policies for deadlock avoidance.

6. CONCLUSION

In this paper a deadlock problem in a class of PN models has been discussed. This class of PN represents FMS with multiple (one-unite-of-resource-per-operation) production sequences. For this class of systems, an on-line deadlock avoidance algorithm based on the structure of the PN model (and not on its reachability graph) has been presented. This algorithm has been shown to be less restrictive than the DAA proposed in [3]. In fact, it has been shown that any DAA for the PN class studied in this paper is at least as restrictive as DAA α stated in section 3. The proposed DAA has been applied to the problem of allocating finite buffer space in a multicell machining facility. Other potential applications, such as multirobot assembly cell, and coordination of AGVs on a shop floor, are described in [3].

Resource allocation problems in FMS with multiple nonsequential production processes (i.e., production processes which include branching and/or merging operations) with multiple-resources- per-operation are known to be hard. It is therefore important to identify classes of PN models for which necessary and sufficient conditions for deadlock avoidance can be derived in polynomial time. Extensions of the approach taken in this work to larger classes of PN models, as well as to dispatching strategies that issue simultaneous production actions, are topics for further research.

REFERENCES

[1] A. Burns and A. Wellings. *Real-time Systems and their Programming Languages*. Addison-Wesley, 1990.

[2] A.N. Habermann. Prevention of systems deadlocks. *Communications of the ACM*, 12(7):373–385, July 1969.

[3] Z.B. Banaszak and B.H. Krogh. Deadlock avoidance in flexible manufacturing systems with concurrently competing process flows. *IEEE Trans. on Robotics and Automation*, 6(6):724–734, December 1990.

[4] E. Roszkowska and Z. Banaszak. Problems of deadlock handling in pipeline processes. *Computer and Information Sciences VI*, pages 1185–1194, 1991.

[5] M. Silva and R. Valette. Petri nets and flexible manufacturing. *Lecture Notes in Computer Science*, 424:374–417, 1989.

[6] Y. Brave and D. Bonvin. A minimally restrictive policy for deadlock avoidance in a class of fms. Technical Report # 1992.03, Institut d'Automatique, Ecole Polytechnique Fédérale de Lausanne (EPFL), Lausanne, Suisse, May 1992.

[7] J. L. Peterson. *Petri net theory and the modeling of systems*. Prentice-Hall, Englewood Cliffs, NJ, 1981.

SIMILARITY OF EVENTS IN DISCRETE EVENT SYSTEMS

Jiří Pik*

Abstract. In the paper, the similarity of events in the discrete event systems is proposed. Using this, the event hierarchy and the event uncertainty are considered. Following a hierarchical event structure, a concept of macro-events is introduced. To characterise the used formalism, algebraic linguistics and structural pattern recognition techniques are utilized, especially those of string matching. A simple example is included to illustrate the developed ideas in a real application.

1. INTRODUCTION

Although different types of uncertainty can occur in the analysis, modelling and control of discrete event systems, no uncertainty has been supposed in the pioneering work of Ramadge and Wonham [1] and [2], and in much of the work that followed. To consider some real situations, partial observations of the process events specified by a mask or observation function have been introduced in [3]. Lin and Wonham incorporated these results into the decentralised supervision structure [4], and recent work [5], concerns with uncertain transitions in the finite automaton model of discrete-event systems.

To deal with uncertainties in actual discrete event systems, we are faced to handle noisy, distorted, or incomplete data. As this problem is very closely related to similar problems of structural pattern recognition or processing of non-numerical data in general, we believe it will be useful to adopt some methods and techniques from these fields for discrete event systems.

2. BASIC CONCEPTS

An alphabet is a finite nonempty set the elements of which we call event symbols. If E is an alphabet, then E^* denotes the set of all strings or words of finite length composed of the event symbols from the alphabet E including the string ϵ consisting of no symbols. The set $E^* - \{\epsilon\}$ is denoted by E^+.

The length of a string X, written $|X|$, means the number of symbols in X when each symbol is counted as many times as it occurs, $|\epsilon| = 0$. The set of all strings X over E with $|X| \leq m$, $m > 0$, is denoted by E^{*m}. The set $E^{*m} - \{\epsilon\}$ is denoted by E^{+m}.

A string X is a substring of a string Y iff there are strings X_1 and X_2 such that $Y = X_1 X X_2$, where $X_1 X X_2$ denotes the concatenation of the strings X_1,

*Czechoslovak Academy of Sciences, Institute of Information Theory and Automation, Pod vodárenskou věží 4, 182 08 Prague 8, Czechoslovakia; e-mail: pik@cspgas11.bitnet
The work was partially supported by the grants ČSAV #27502 and ČSAV #27558.

X, and X_2. The string ϵ is an identity element in E^*, thus $\epsilon X = X\epsilon = X$ for all $X \in E^*$.

Let φ be an equivalence relation over a set T. Then for every $t \in T$, the set $B_\varphi(t) = \{u : t\varphi u\}$ is an equivalence class. A partition π on T is a collection of disjoint subsets of T, called blocks of π, whose set union is T.

3. EVENT SIMILARITY

Different approaches ranging from pure experimental to more theoretically based ones can be considered to introduce the notion of the similarity of events. In what follows, the Levenshtein metric [6], and the corresponding weighted distance that can be interpreted as a similarity measure of the strings of the event symbols are considered.

Using a-priori defined symbol-to-symbol operations, the strings of the event symbols are transformed into new ones.

Let E be an alphabet, an operation over E is an ordered pair $s = (a, b)$ such that $a, b \in E \cup \{\epsilon\}$ and $s \neq (\epsilon, \epsilon)$. The operation $s = (a, b)$ is called

1. deletion if $b = \epsilon$,

2. insertion if $a = \epsilon$,

3. substitution otherwise.

A string Y results from the application of the operation $s = (a, b)$ to a string X, written $X \overset{s}{\Longrightarrow} Y$, if $X = A_1 a A_2$ and $Y = A_1 b A_2$, where $A_1, A_2 \in E^*$. To transform a string X into Y, a sequence of operations $S = s_1, s_2, \ldots, s_q$ is needed, $q > 0$, such that $X = X_0 \overset{s_1}{\Longrightarrow} X_1, \; X_1 \overset{s_2}{\Longrightarrow} X_2, \ldots, X_{q-1} \overset{s_q}{\Longrightarrow} X_q = Y$.

Example 1 *Let us consider* $E = \{a, b, c\}$, *let* $X = abcac$ *and* $Y = acbcba$. *The application of the following sequence of operations* $S_1 = s_1, s_2, s_3$ *transforms* X *into* Y

$$
\begin{array}{lll}
X = X_0 = abcac & s_1 = (\epsilon, c) & X_1 = acbcac \\
& s_2 = (\epsilon, b) & X_2 = acbcbac \\
& s_3 = (c, \epsilon) & X_3 = acbcba = Y.
\end{array}
$$

As an alternative to S_1, *the sequence* $S_2 = s_1, s_2, s_3, s_4, s_5$ *can be considered*

$$
\begin{array}{lll}
X = X_0 = abcac & s_1 = (b, c) & X_1 = accac \\
& s_2 = (\epsilon, b) & X_2 = acbcac \\
& s_3 = (a, \epsilon) & X_3 = acbcc \\
& s_4 = (\epsilon, b) & X_4 = acbcbc \\
& s_5 = (c, a) & X_5 = acbcba = Y.
\end{array}
$$

To reflect a difference in the application of the operations, a real number $w(s)$ called a weight of the operation $s = (a, b)$ is associated with each operation. The following properties are required for all $a, b, c \in E \cup \{\epsilon\}$

(i) $w(a, b) \geq 0$ and $w(a, b) = 0$ iff $a = b$,

(ii) $w(a, b) = w(b, a)$,

(iii) $w(a, b) + w(b, c) \geq w(a, c)$.

The notion of $w(a, b)$ is extended to a sequence of operations $S = s_1, s_2 \ldots, s_q$ by

$$w(S) = \sum_{i=1}^{q} w(s_i) \text{ and by } w(S) = 0 \text{ for } q = 0.$$

The weighted distance $d_w(X, Y)$ between the strings X and Y is defined by

$$d_w(X, Y) = \min\{ w(S) : S \text{ is a sequence of operations transforming } X \text{ into } Y \}.$$

Procedures following the algorithm of Wagner and Fischer [7], are commonly used to determine the distance. While the original algorithm [7] takes $O(|X||Y|)$ time and space, the algorithms [8, 9, 10] for a slightly reduced problem take $O(|X| + |Y|)$.

Example 2 *Let us complete Example 1 by letting* $w(\epsilon, x) = 0.5$, $w(x, \epsilon) = 0.5$, *and* $w(x, y) = 0.25$ *for all* $x, y \in \{a, b, c\}$. *Summing* $w(\cdot, \cdot)$ *for* S_1 *and* S_2 *we get* $w(S_1) = 0.5 + 0.5 + 0.5 = 1.5$ *and* $w(S_2) = 0.25 + 0.5 + 0.5 + 0.5 + 0.25 = 2.0$. *Finally, to determine the weighted distance between* X *and* Y, *a sequence* $S_d = s_1, s_2, s_3$ *with* $w(S_d) = \min\{w(S)\}$ *is found*

$$
\begin{array}{lll}
X = X_0 = abcac & s_1 = (\epsilon, c) & X_1 = acbcac \\
 & s_2 = (a, b) & X_2 = acbcbc \\
 & s_3 = (c, a) & X_3 = acbcba = Y.
\end{array}
$$

As $w(S_d) = 0.5 + 0.25 + 0.25 = 1$, *the weighted distance* $d_w(X, Y) = 1$.

A graphical interpretation of the algorithm [7] is depicted in figure 1. The considered symbol-to-symbol operations, namely deletion, insertion and substitution, correspond to vertical, horizontal and diagonal branches of the lattice of figure 1, respectively. The weight of the corresponding operation is associated with every branch. In the lattice, each oriented path between the circled vertices represents a sequence of operations S transforming X into Y. The sum of weights along such a path is $w(S)$ and to determine $d_w(X, Y)$, the path corresponding to a minimal value of $w(S)$ is found.

Continuing the examples, the path representing the sequence $S_e = (x_1, y_1), (\epsilon, y_2), (x_2, y_3), (x_3, y_4), (x_4, y_5), (x_5, y_6)$ is emphasized in figure 1. If $X = abcac$ and $Y = acbcba$, then $w(S_e) = w(S_d)$.

Example 3 *Let us consider a string* $Z = acbac$ *and compare the weighted distance between* X *and* Z *and that between* Y *and* Z. *Supposing* $w(\cdot, \cdot)$ *as given above, we get* $d_w(X, Z) < d_w(Y, Z)$ *and conclude that* $X = abcac$ *is more similar to* Z *than* $Y = acbcba$.

A class of similarity measures instead of a single one is obtained through the specification of the weights of the operations. The proper choice of the weight values usually reflects our knowledge and insights into the considered problem. Another approach based on a weight parametrisation and a given sample set of strings is proposed in [11].

A probabilistic interpretation of the introduced distance is possible [12], where the weighted distance $d_w(X, Y)$ corresponds to the most likely way of transforming X into Y. Further, a close relationship can be found between the weighted distance $d_w(X, Y)$ and some similarity coefficients introduced experimentally [13].

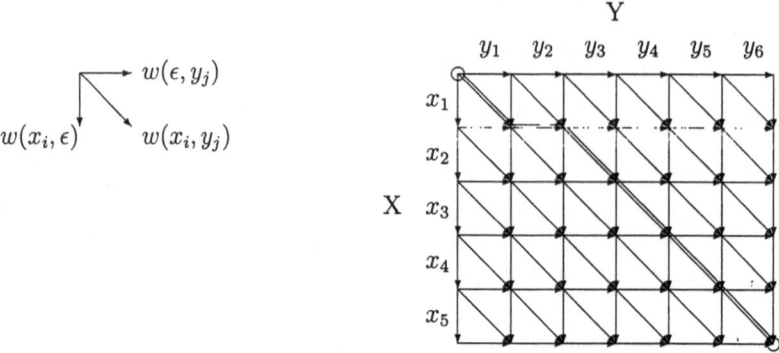

Figure 1: The lattice for $X = x_1 \ldots x_5$ and $Y = y_1 \ldots y_6$.

An extension of the notion of the symbol-to-symbol operation is possible, for instance, a context sensitive application of the operations or a generalised form introducing string-to-symbol and string-to-string operations, e.g. [14].

4. EVENT HIERARCHY

An equivalence relation over the set of the strings of the event symbols and an induced partition of this set can be defined using the weighted distance. The strings belonging to the same partition block can be represented by a macro-event symbol and a new representation of each string based on these symbols is possible. Hence, the similarity of the events can be viewed as a basis of an event hierarchy, cf. e.g. [15], where the frequency with which the actual events occur during the production process plays an analogous role.

A clustering of strings based on the nearest neighbour rule [16], can be utilized to define an alternative equivalence relation.

A difficult problem can arise if the considered set of the strings is not presented. As the hierarchical representation is based on an idea of a sample path as a composition of separate parts, difficulties and ambiguities are met in an inverse problem: a segmentation of the given sample path into substrings. In this case, additional information is needed like a limited length or location of the substrings, occurrences of distinguished or marked event symbols, etc.

From another point of view, the similarity of the event symbols provides a general framework to apply the selectivity principle that belongs to the funda-mental principles of human reasoning and problem solving. Using this principle, one can abstract from details of a process description and work in terms of the simpler modules or abstractions.

Following the foregoing considerations, our approach is to define a hierarchical structure of the set of the strings of the event symbols based on the notin of the event similarity.

A hierarchical event structure, HES, is defined as a 3-tuple $HES = (h, E, F)$, where

h is a number of the levels of hierarchy , $h > 1$,

E is an alphabet of HES, $E = \bigcup\limits_{i=1}^{h} E_i$,

    the alphabets E_i, E_j are not necessarily disjoint for $i, j = 1, \ldots, h$,

F is a set of relations $F = \{\varphi_2, \ldots, \varphi_h\}$,

and for each $i = 2, \ldots, h$:

(i) there exist $m_i, n_i > 0$, $n_i < i$, such that the relation φ_i is an equivalence over $E^{*m_i}_{n_i}$,

(ii) there exists a mapping ψ_i from the induced partition $\pi_i = \{B_{n_i,1}, B_{n_i,2}, \ldots\}$ on $E^{*m_i}_{n_i}$ onto $E_i \cup \{\epsilon\}$,

(iii) the symbols of the alphabet E_i are called macro-event symbols (with respect to the n_i-th hierarchical level).

To define a hierarchy of the strings, let $HES = (h, E, F)$ be a hierarchical event structure, let for some $i = 2, \ldots, h$ there exist $X \in E^+_{n_i}$ and $Y \in E^+_i$. Then define Y to be a direct representation of X, written $X \overset{HES}{\Longrightarrow} Y$, if the following conditions are satisfied:
$X = X_1 \ldots X_k \ldots X_l$, and $Y = y_1 \ldots y_k \ldots y_l$, $l > 0$, where $X_k \in E^{*m_i}_{n_i}, y_k \in E_i \cup \{\epsilon\}$, such that for each $k = 1, \ldots, l$ there exists a block $B_{n_i,j}$ of π_i on $E^{*m_i}_{n_i}$ for that $X_k \in B_{n_i,j}$ and $y_k = \psi_i(B_{n_i,j})$.

Let $X \in E^+_j$ and $Y \in E^+_i$ for $1 \leq j < i \leq h$. Then Y is called a representation of X (on the i-th hierarchical level) if there exist $z > 0$ and a sequence $Z_0, \ldots, Z_z$, such that $Z_0 = X$, $Z_z = Y$, and $Z_{t-1} \overset{HES}{\Longrightarrow} Z_t$ for each $t = 1, \ldots, z$.

A tree of the levels of hierarchy is given through the pairs (n_i, i), the hierarchy is called linear if $n_i = i - 1$ for each $i = 2, \ldots, h$.

5.   EVENT UNCERTAINTY

Two categories of event uncertainties can basically occur in some real applications of the discrete event systems, namely the category of uncertainties that are due to possibly ambiguous event recognition or description and uncertainties owing to the transmission of information through a noisy channel. As a result the controller or supervisor $\mathcal{S}$ observes an event string that is different from an event string of the controlled system or generator $\mathcal{G}$. The differences we distinguish are:

1. an actual event is not observed by $\mathcal{S}$,
2. an observed event is not in the string of $\mathcal{G}$,
3. an actual event of $\mathcal{G}$ is observed by $\mathcal{S}$ as a different event.

Obviously, these differences can be modelled using event deletions, insertions or substitutions, and, therefore, the operations $s_1, s_2, \ldots, s_q$ are considered to transform the event string X of $\mathcal{G}$ into Y observed by $\mathcal{S}$.

Let $E_{\mathcal{G}}$ and $E_{\mathcal{S}}$ denote the alphabets of $\mathcal{G}$ and $\mathcal{S}$, respectively. Let m denotes a length of the considered substrings. Formally, a general deformation model of the strings of the event symbols can be introduced by $HES = (h, E, F)$, where

$$\begin{aligned} h &= 2, \\ E &= E_{\mathcal{G}} \cup E_{\mathcal{S}}, \\ F &= \{\varphi_2\}, \end{aligned}$$

and

(i) the relation φ_2 is an equivalence over $E_{\mathcal{G}}^{*m}$ de fined using the weighted distance based on the considered operations,

(ii) the mapping ψ_2 is a correspondence between partition blocks of π_2 on $E_{\mathcal{G}}^{*m}$ and the symbols of $E_{\mathcal{S}} \cup \{\epsilon\}$,

(iii) the macro-event symbols of $E_{\mathcal{S}}$ represent both single events and abstractions from them including differences due to uncertainty that occur in strings over $E_{\mathcal{G}}$.

Clearly, if $m = 1$ then an equivalence φ_2 over $E_{\mathcal{G}} \cup \{\epsilon\}$ and the weighted distance reduced to the single weights $w(s)$ are considered.

6. APPLICATION

The similarity of events presented in this paper can be illustrated by a simple example based on a real application of the developed ideas in the method of analysis and modelling of the atmospherical circulation. This method is a part of the computer-aided decision support system for the long-range weather forecasting in the Czech Institute of Hydrometeorology.

In the application, the updated standardisation (GWL event symbols) of the pressure field over the Atlantic-Europe region is considered [17]. Starting from 1881, 29 well defined non-numerical types of the daily configurations of the pressure fields are distinguished. Moreover, an exceptional configuration is represented by another type. The corresponding alphabet of GWL symbols is $E_{GWL} = \{C, S, W, A, B, H, V, X, Z, Y, T, R, J, I, F, E, M, O, N, D, 1, 2, 3, 4, 5, 6, 7, 8, 9, U\}$.

There are attempts to utilize this time series, containing about 40,000 symbols and constituting a sample path, to the weather forecasting; e.g. [18], where the Markov chain theory is taken into consideration. The obtained results are not, however, significant for the forecasting.

In our approach, the set of the strings of the limited length over the alphabet E_{GWL} is reduced using a hierarchical event structure. From the meteorological point of view, the considered hierarchical event structure introduces a new standardisation derived from the original one and based on the macro-event symbols. To get the considered set of the strings, substrings with properties reflecting some meteorological parameters (e.g., the length, time location, trends, etc.) are extracted from the given sample path.

A two-level hierarchy is defined by $HES = (h, E, S)$, where

$$\begin{aligned} h &= 2, \\ E &= E_{GWL} \cup E_M, \\ F &= \{\varphi_2\}. \end{aligned}$$

The limited length of the substrings over G_{GWL} is denoted by m and the alphabet of macro-event symbols by E_M.

The equivalence φ_2 over E_{GWL}^{*m} is defined using the weights $w(s)$ of the symbol-to-symbol operations; the mapping ψ_2 is a correspondence between blocks of the partition on E_{GWL}^{*m} and the symbols of $E_M \cup \{\epsilon\}$.

To define φ_2, two sets of the weights of the operations are taken into account. The weights $w(s) = w(a,b)$, $a,b \in E_{GWL} \cup \{\epsilon\}$, of the former set follow a physical analogy of the pressure fields and are as follows

$$w(a,a) = 0 \quad \text{for all } a \in E_{GWL},$$
$$w(a,b) \leq c_1 \quad \text{iff } a,b \in \sigma_i \text{ for some } i \in \{1,2,\ldots,10\},\ a \neq b,$$
$$w(a,b) \geq c_2 \quad \text{iff } a \in \sigma_i,\ b \in \sigma_j\ i,j \in \{0,1,\ldots,10\} \text{ and } i \neq j,\ 0 < c_1 < c_2,$$

where

$$\sigma_0 = \{\epsilon\},\ \sigma_1 = \{C,S,W\},\ \sigma_2 = \{A,B\},\ \sigma_3 = \{H\},$$
$$\sigma_4 = \{V,X,Z,Y,T,R\},\ \sigma_5 = \{J,I\},\ \sigma_6 = \{F,E,M,O\},$$
$$\sigma_7 = \{N,D\},\ \sigma_8 = \{1,2,3,4,7,8,9\},\ \sigma_9 = \{5,6\},\ \sigma_{10} = \{U\},$$

and

$$\begin{aligned}
E_M &= \{\mu_1,\ldots,\mu_{10}\}, \\
\psi_2(\epsilon) &= \epsilon, \\
\psi_2(\sigma_i^{+m}) &= \mu_i \text{ for } i = 1,\ldots,10.
\end{aligned}$$

The latter set of the weights represents a more complex case as the weights depend on the correspondence between the weather and the standardisation and, in effect, the twelve sets of the monthly weights are introduced.

REFERENCES

[1] P. J. Ramadge and W. M. Wonham. Supervisory Control of a Class of Discrete-Event Processes. *SIAM J. Control and Optimization*, Vol.25, No.1, 206–230, 1987.

[2] W. M. Wonham and P. J. Ramadge. On the Supremal Contr ollable Sublanguage of a Given Language. *SIAM J. Control and Optimization*, Vol.25, No.3, 637–659, 1987.

[3] R. Cieslak, C. Desclaux, A. S. Fawaz and P. P. Varaiya. Supervisory Control of Discrete-Event Processes with Partial Observations. *IEEE Trans. on Automatic Control*, AC–33, No.3, 249–260, 1988.

[4] F. Lin and W. M. Wonham. Decentralized Control and Cooperation of Discrete-Event Systems with Partial Observation. *IEEE Trans. on Automatic Control*, Vol. AC-35, No.12, 1330–1337, 1990.

[5] S. Young and V. K. Garg. Transition Uncertainty in Discrete Event Systems. *In: Workshop on Discrete Event Systems*. June, 21–23, 1991, Amherst, U.S.A., 1991.

[6] V. I. Levenshtein. Binary Codes Capable of Correcting Deletions, Insertions and Reversals. *Sov. Phys. Dokl.*, Vol.10, No.8, 707–710, Feb., 1966.

[7] R. A. Wagner and M. J. Fischer. The String-to-String Correction Problem. *J.ACM*, Vol.21, No.1, 168–173, 1974.

[8] J. Chang, O. Ibarra and M. Palis. Parallel Parsing on a One-Way Array of Finite-State Machines. *Trans. Comput.*, Vol. C-36, No. 1, pp.64–75, 1987.

[9] O. H. Ibarra, T. Jiang and H. Wang. String Editing on a One-Way Linear Array of Finite-State Machines. *IEEE Trans. Comput.*, Vol. C-41, No. 1, pp. 112-118, 1992.

[10] W. J. Masek and M. S. Paterson. A Faster Algorithm Computing String Edit Distance. *Journal of Computer and System Sciences*, Vol. 20, pp. 18–31, 1980.

[11] H. Bunke and J. Csirik. Inference of Edit Costs using Parametric String Matching. *Proc. of the 11th IAPR Int. Conf. on Pattern Recognition*, The Hague, Vol.2, 549-552, 1992.

[12] S. Y. Lu and K. S. Fu. Stochastic Error-Correcting Syntax Analysis for Recognition of Noisy Patterns. *IEEE Trans. on Comp.*, Vol. C-26, No.12, 1268–1276, 1977.

[13] J. Pik. Structural Analysis of Experimental Curves in Numerical Taxonomy. *Proc. of the 8th IAPR Int. Conf. on Pattern Recognition*, Paris, Vol.1, 389–391, 1986.

[14] K. Abe and N. Sugita. Distances between Strings of Symbols – Review and Remarks. *Proc. of the 6th IAPR Int. Conf. on Pattern Recognition*, Munich, Vol.1, 172–174, 1982.

[15] S. B. Gershwin. Hierarchical Flow Control: A Framework for Scheduling and Planning Discrete Events in Manufacturing Systems. *Proceedings of the IEEE*, Vol.77, No.1, 196–209, 1989.

[16] S. Y. Lu and K. S. Fu A Sentence-to-Sentence Clustering Procedure for Pattern Analysis. *IEEE Trans. on Systems, Man, and Cybernetics*, Vol. SMC-8, No.5, 381–389, 1978.

[17] P. Hess and H. Brezowsky. Katalog der Grosswetterlagen Europas. *Deutsch. Wetterd.*, Offenbach a.M., 1969.

[18] C. Mares and I. Mares. Testing of the Markov dependence for certain meteorological parameters. *Ninth Prague Conf. on Inf. Theory, Stat. Dec. Functions and Random Proc.*, Prague, 1982.

CONTROL OF DISCRETE EVENT SYSTEMS BY MEANS OF THE BOOLEAN DIFFERENTIAL CALCULUS

Rainer Scheuring   Hans Wehlan *

Abstract.   A new approach to the study of discrete event systems (DES), characterized by automata, Petri-Nets or related presentations, is proposed. The Boolean Differential Calculus (BDC) supports modeling, analysis and synthesis of DES. This paper not only demonstrates fundamental properties of the BDC, but also presents a synthesis algorithm for the cat-mouse-example.

1. INTRODUCTION

Logical, algebraic, and performance oriented models have been developed for DES. Logical models are principally intended for the study of the qualitative logical aspects of the system dynamics (e.g. correct use of resources, coordination of concurrent processes). Examples of logical models include: finite state machines, formal languages and Petri-Nets. Timed models incorporate time as an integral part of the model structure and are especially suited for answering performance-related questions [1].

The description of DES by means of the Boolean Differential Calculus (BDC) basically belongs to the class of logical untimed models, but the BDC also provides features that help to improve the handling of timed logical models. The BDC enables us to work with automata and Petri-Nets in a common framework [2]. As a consequence, concepts and algorithms of automata- [3], [4], [5], Petri-Net- [6] and supervisory-control-theory [7], [8], [9], [10], [11], [12] can be integrated and applied to DES.

The BDC was initially developed in 1959 with the papers of Akers [13] and Talantsev [14]. Since then considerable work has been done in both the theory and application of this method mainly in the area of digital system design. The work of Bochmann and Posthoff [15], Thayse [16] and many others established the BDC as a mathematical theory that shows deep relations with continuous variable system theory. For a comprehensive presentation of the BDC, see Bochmann and Posthoff [15].

Although the combinatorial explosion of states in 'big' systems is significantly reduced through a modular and hierarchical approach, not too many results can be obtained without software tools. The software system XBOOLE [17] and its restricted public-domain-version XBMini [18] are able to calculate all operations of the BDC for rather large systems. XBOOLE provides the user with a C routine

*Institut für Systemdynamik und Regelungstechnik, Universität Stuttgart, Pfaffenwaldring 9, 7000 Stuttgart 80, Germany; e-mail: rs@isr.verfahrenstechnik.uni-stuttgart.dbp.de

library that contains about 100 functions. These functions perform operations on solution sets of Boolean equations. The solution sets are represented with ternary matrices whose elements are 1, 0 and $-$ (don't care).

In this paper, we will concentrate on basic numerical aspects of the BDC and the computer program XBOOLE. Although these aspects can be hidden behind a comfortable user interface that uses symbolic variables and graphical representations, we will choose a tedious representation with tables of 1, 0 and don't care elements to provide some insight into the underlying numerics.

The paper is divided into three main sections. Section 2 is a survey of the BDC. It is intended to be introductory, but with sufficient details to support the understanding of Section 3. Even though the BDC has been successfully applied to realistic process control problems [19], Section 3 proposes a synthesis algorithm for the cat-mouse-example [20] that enables a comparison of this approach with other ones. In Section 4 we summarize our results.

2. BOOLEAN DIFFERENTIAL CALCULUS (BDC)

2.1 Standard Model

A discrete event system (DES) is modeled by a generalized automaton

$$G = (X, U, Y, \delta, \lambda) \tag{1}$$

where X is the set of states; U is the input alphabet; Y is the output alphabet; $\delta : X \times U \longrightarrow X$ is the transition function; $\lambda : X \times U \longrightarrow Y$ is the output function. X, U and Y are finite sets. The DES (1) may have not only one but many active states at one moment in time. As a consequence, nondeterministic automata as well as Petri-Nets can be easily transformed into this representation [2].

The Boolean Differential Calculus (BDC) also describes DES with the following standard model that consists of Boolean equations [2], [21]

$$^1\mathbf{x} = \mathbf{f}(\mathbf{x}, \mathbf{u}) \tag{2}$$
$$0 = \mathbf{g}(\mathbf{x}, \mathbf{u}) \tag{3}$$
$$\mathbf{y} = \mathbf{h}(\mathbf{x}, \mathbf{u}), \tag{4}$$

where $\mathbf{u}$, $\mathbf{x}$, $\mathbf{y}$ are Boolean input, state and output vectors, $^1\mathbf{x}$ denotes the next or following state vector. $\mathbf{f}$, $\mathbf{g}$, $\mathbf{h}$ are Boolean functions of appropriate dimensions, which carry out the logical operators $\wedge$ (AND) [1], $\vee$ (OR), $^-$(NOT) and $\oplus$ (XOR) on $\mathbf{u}$, $\mathbf{x}$, $^1\mathbf{x}$ and $\mathbf{y}$. Just as the Boolean equation (2) represents the dynamical behaviour of DES, so is equation (3) useful to express static conditions.

Using $a = b \Leftrightarrow 0 = a \oplus b$ and $(a = 0)(b = 0) \Leftrightarrow a \vee b = 0$, we find equivalent representations for the equations (2), (3) and (4):

$$\begin{aligned}
^1\mathbf{x} &= \mathbf{f}(\mathbf{x}, \mathbf{u}) \\
\Leftrightarrow \quad 0 &= {}^1\mathbf{x} \oplus \mathbf{f}(\mathbf{x}, \mathbf{u}) \\
\Leftrightarrow \quad 0 &= \bigvee_{i=1}^{n} {}^1x_i \oplus f_i(\mathbf{x}, \mathbf{u}),
\end{aligned} \tag{5}$$

---

[1]The $\wedge$-operator is omitted in this paper.

$$0 = \bigvee_{i=1}^{o} g_i(\mathbf{x}, \mathbf{u}), \tag{6}$$

$$0 = \bigvee_{i=1}^{l} y_i \oplus h_i(\mathbf{x}, \mathbf{u}), \tag{7}$$

where n and l are equal to the length of the vectors $\mathbf{x}$ and $\mathbf{y}$ and o is equal to the number of static boundary conditions. With the transition and output functions

$$\delta(^1\mathbf{x}, \mathbf{x}, \mathbf{u}) = \bigvee_{i=1}^{n} {}^1x_i \oplus f_i(\mathbf{x}, \mathbf{u}) \vee \bigvee_{i=1}^{o} g_i(\mathbf{x}, \mathbf{u}) \tag{8}$$

$$\lambda(\mathbf{y}, \mathbf{x}, \mathbf{u}) = \bigvee_{i=1}^{l} y_i \oplus h_i(\mathbf{x}, \mathbf{u}) \tag{9}$$

we obtain the Boolean equation

$$0 = \omega(\mathbf{y}, {}^1\mathbf{x}, \mathbf{x}, \mathbf{u}) = \delta(^1\mathbf{x}, \mathbf{x}, \mathbf{u}) \vee \lambda(\mathbf{y}, \mathbf{x}, \mathbf{u}). \tag{10}$$

Analogous to the 'Principle of Least Action', a DES can change its state, if $\omega(\mathbf{y}, {}^1\mathbf{x}, \mathbf{x}, \mathbf{u}) = 0$. Therefore the function ω is called the Boolean Lagrangian function [15].

For numerical reasons it has been proven to be efficient to work with sets of solutions of boolean equations [17]:

Definition 1 *A set Ω is called the solution set of the Boolean equation*

$$0 = \omega(\mathbf{y}, {}^1\mathbf{x}, \mathbf{x}, \mathbf{u}), \tag{11}$$

if and only if Ω contains all vectors $[\mathbf{y} {}^1\mathbf{x} \mathbf{x} \mathbf{u}]$ that solve equation (11) [2].

As an example, the Boolean equation

$$0 = a(b \oplus d) \oplus \bar{c} \oplus a(b\bar{c} \oplus d) \tag{12}$$

has the solution set of table 1. The solution is presented in the disjunctive normal form, which means that each row is a conjunction and the union of all rows is the solution. For the symbol '−' (don't care) either 1 or 0 can be inserted. XBOOLE not only takes advantage of the disjunctive but also of the conjunctive, equivalent and antivalent normal forms. The solution set of equation (12) is not calculated by means of a complete enumeration. XBOOLE automatically computes elementary solution sets, for example of terms such as $(b \oplus d)$, and combines these sets with other sets.

In table 1 we find a minimal solution set with an algorithm that is related to the Quine-McCluskey method [22], [23]. Table 2 presents the same solution set in an orthogonal form, this means no solution vector appears in more than one row (compare e. g. $[0\,0\,1\,0]$). Most XBOOLE-operators can be applied to both forms, but the orthogonal one is in many cases more efficient.

---

[2]In this paper we use lower case letters for Boolean functions and upper case letters for their solution sets.

a	b	c	d
0	-	1	-
-	0	1	-

Table 1: Solution set of equation (12).

a	b	c	d
0	1	1	-
-	0	1	-

Table 2: Orthogonal solution set of equation (12).

Cat-mouse example: Wonham and Ramadge [20] proposed the following example, where a cat and a mouse are placed in a maze (see figure 1, [24]). Initially the cat is in room 2 and the mouse is in room 4. Each door in the maze can be traversed only by the mouse and by the cat, respectively, in the direction indicated. In addition each door, with the exception of c_7, can be opened or closed by means of control actions.

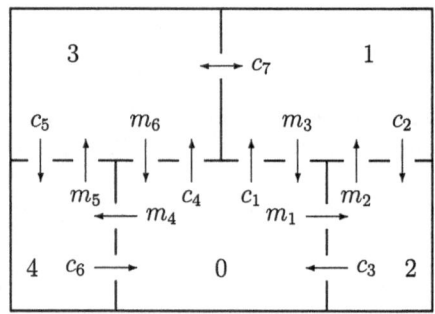

Figure 1: Maze of the cat-mouse-example.

The objective is to find a DES-controller that permits the cat and the mouse the greatest possible freedom of movement but also guarantees that

- the animals never occupy the same room simultaneously;

- it is always possible for the cat and the mouse to return to the initial position (cat: room 2; mouse: room 4).

The movement of the cat and the mouse in the maze may be described with automata; one automaton for the cat and one for the mouse (figure 2). In the BDC every input, state or output has to be typified with a binary code. As an example, we introduce the map of table 3 that represents the states of the two automata. For the inputs we have

$$\mathbf{u}_m = [m_1 \ m_2 \ m_3 \ m_4 \ m_5 \ m_6] \qquad (13)$$
$$\mathbf{u}_c = [c_1 \ c_2 \ c_3 \ c_4 \ c_5 \ c_6 \ c_7], \qquad (14)$$

as can be seen in table 4 for the cat.

room	x_{i1}	x_{i2}	x_{i3}
0	0	0	0
1	1	0	0
2	0	1	0
3	1	1	0
4	0	0	1

Table 3: State space representation of the rooms, where $i \in \{c, m\}$.

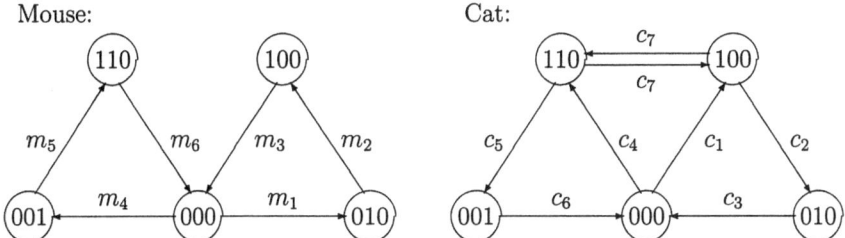

Figure 2: Automata for the cat and the mouse.

When we work with automata, the derivation of the solution set Ω (see Definition 1) is trivial. Using a state diagram or a transition table of an automaton, we can immediately write down its solution set Ω. In table 4 we find the solution set of the cat for the automaton of figure 2.

x_{c1}	x_{c2}	x_{c3}	$^1x_{c1}$	$^1x_{c2}$	$^1x_{c3}$	c_1	c_2	c_3	c_4	c_5	c_6	c_7
0	0	0	0	0	0	0	0	0	0	0	0	0
1	0	0	1	0	0	0	0	0	0	0	0	0
0	1	0	0	1	0	0	0	0	0	0	0	0
1	1	0	1	1	0	0	0	0	0	0	0	0
0	0	1	0	0	1	0	0	0	0	0	0	0
0	0	0	1	0	0	1	0	0	0	0	0	0
1	0	0	0	1	0	0	1	0	0	0	0	0
0	1	0	0	0	0	0	0	1	0	0	0	0
0	0	0	1	1	0	0	0	0	1	0	0	0
1	1	0	0	0	1	0	0	0	0	1	0	0
0	0	1	0	0	0	0	0	0	0	0	1	0
1	0	0	1	1	0	0	0	0	0	0	0	1
1	1	0	1	0	0	0	0	0	0	0	0	1

Table 4: Solution set Ω_c for the Boolean Lagrangian of the cat ω_c.

On the other hand, for a BDC-model of a Petri-Net, we have to formulate a Boolean equation of type (2) for each state of the Net. As an example, considering the input and output transitions of x_3 in figure 3, we obtain

$$^1x_3 = x_3 \oplus (\bar{x}_3 x_1 u_1 \vee \bar{x}_3 x_2 u_2 \vee x_3 \bar{x}_4 \bar{x}_5 u_3). \tag{15}$$

If u_1 and u_2 are supposed not to be equal to 1 at the same instant, this restriction can be satisfied with

$$0 = u_1 u_2. \tag{16}$$

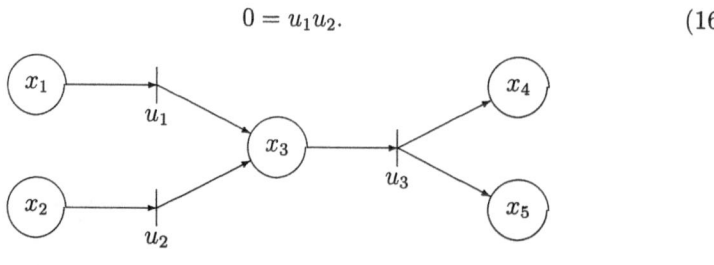

Figure 3: Petri-Net with places x_1 - x_5 and transitions u_1 - u_3.

Continuing this procedure and calculating the transition and output functions δ and λ (equations (8) and (9)), we get the Boolean Lagrangian ω (equation (10)) of the Petri-Net. As far as Petri-Nets are especially suited to model parallel and concurrent processes [25], we see that the BDC is able to represent such processes in a natural way.

In the cat-mouse-example it is appropriate to start with the Lagrangians ω_m for the mouse and ω_c for the cat. In Section 3 we develop a discrete event controller that can be written as $0 = g_{con}(\mathbf{y}, \mathbf{u})$ [3]. As a result, we get the function

$$\omega = \omega_m \vee \omega_c \vee g_{con}, \tag{17}$$

which specifies the behaviour of the overall system. In addition, the cardinality of the set

$$\Omega = \Omega_m \cap \Omega_c \cap G_{con} \tag{18}$$

is substantially reduced through the controller g_{con}. Almost always in the design of discrete event controllers for chemical processes we have experienced the fact that the combinatorial explosion of states can be eliminated with a proper design of discrete event controllers.

2.2 Boolean Differential Operators

A group of more than twenty Boolean differential operators takes the key position in the derivation of valuable information about DES. Many of these operators are Boolean functions of $f(\mathbf{a})$ and $\partial f(\mathbf{a})/\partial a_i$.

Definition 2 (Bochmann and Posthoff [15])
If $f(a_1, \ldots, a_i, \ldots, a_n)$ is a Boolean function on the n Boolean variables $a_1, \ldots, a_n$, the partial derivative of $f(\mathbf{a})$ with respect to a_i is defined by

$$\frac{\partial f(\mathbf{a})}{\partial a_i} = f(a_1, \ldots, a_i, \ldots, a_n) \oplus f(a_1, \ldots, \bar{a}_i, \ldots, a_n). \tag{19}$$

---

[3]In Section 3 we find that a static output feedback is sufficient. Therefore, we label the controller with g_{con} instead of ω_{con} (see also equation (3)).

For instance, the partial derivative $\partial f(\mathbf{a})/\partial a_1$ of $f(\mathbf{a}) = (a_1 a_2 \vee a_1 a_3)$ yields

$$\frac{\partial(a_1 a_2 \vee a_1 a_3)}{\partial a_1} = a_2 \vee a_3. \tag{20}$$

If a_2 or a_3 are equal to 1 and if a_1 changes its value from 0 to 1 or vice versa, then the function $f(\mathbf{a})$ changes its value, too.

In this paper the min- and max-operators

$$\min_{a_i} f(\mathbf{a}) = f(a_1, \ldots, a_i, \ldots, a_n)\, f(a_1, \ldots, \bar{a}_i, \ldots, a_n) \tag{21}$$

$$\max_{a_i} f(\mathbf{a}) = f(a_1, \ldots, a_i, \ldots, a_n) \vee f(a_1, \ldots, \bar{a}_i, \ldots, a_n) \tag{22}$$

are also used. Differential operators may be applied several times to Boolean functions; e.g. we have

$$\min_{a_{i_1} \cdots a_{i_m}}{}^m f(\mathbf{a}) = \min_{a_{i_1}} \left(\min_{a_{i_2}} \left(\ldots \min_{a_{i_m}} f(\mathbf{a}) \ldots \right) \right). \tag{23}$$

As an example, all states of a DES that have only one successor may be calculated with

$$\wedge_{i=1}^n \frac{\partial}{\partial \dot{x}_i} [\min_{\dot{\mathbf{x}} \backslash \dot{x}_i}{}^{n-1} \omega(\dot{\mathbf{x}}, \mathbf{x})], \tag{24}$$

where

$$\dot{\mathbf{x}} = {}^1 \mathbf{x} \oplus \mathbf{x}. \tag{25}$$

For the automata of the cat and the mouse (figure 2), we find the results of the tables 5 and 6.

x_{c1}	x_{c2}	x_{c3}
0	1	0
0	0	1

x_{m1}	x_{m2}	x_{m3}
1	0	0
0	0	1
-	1	0

Table 5: States with one successor (cat).

Table 6: States with one successor (mouse).

Boolean differential operators provide knowledge about effects of input-, state- or output-changes. Consequently, they offer the opportunity to improve not only the analysis, but also the synthesis of DES.

3. SYNTHESIS ALGORITHM

A calculation of all possible state trajectories of a DES is very helpful in the design and synthesis of DES-controllers. XBOOLE provides the means to calculate these trajectories for rather large systems (see Section 4.1). The trajectories may start with a set X of initial or final (marked) states. To calculate the set of states ${}^1 X$ that can be reached from X in one step, we have to intersect X with the solution set Ω of the Lagrangian of the DES:

$$^1X := \max_{\mathbf{y},\mathbf{x},\mathbf{u}}{}^{l+n+m} (\Omega \cap X). \tag{26}$$

The max-operator is used to remove the variables $\mathbf{y}$, $\mathbf{x}$ and $\mathbf{u}$. The set of all reachable states can easily be calculated, if we perform this step several times. When we start with a final state set and remove the variables $\mathbf{y}$, $^1\mathbf{x}$ and $\mathbf{u}$ in each step, we obtain the coreachable state set. In many cases additional restrictions, e.g. on the set of possible states or inputs, have to be introduced. This can be realized, if we intersect the right side of equation (26) with further solution sets.

Cat-mouse example: Intersecting the initial state of table 7 (step 0) with Ω_c of table 4, we find the two states of step 1 as following states. The continuation of this procedure yields all states that can be reached by the cat. The computational complexity of the intersection does not change when the rows of X contain don't care instead of 1 and 0 elements.

step	x_{c1}	x_{c2}	x_{c3}
0	0	1	0
1	0	-	0
2	-	-	0
3	-	-	0
	0	0	-
4	-	-	0
	0	0	-
5	-	-	0
	0	0	-
⋮		⋮	

Table 7: Reachable states of the cat.

In order to synthesize a discrete event controller for the cat-mouse-problem, we apply the algorithm below, which uses state space information. In our opinion, the exploitation of the state space information is one of the essential aspects of the BDC. In many technical systems discrete states can easily be measured; in addition, control engineers usually find it comfortable to express their knowledge of a system by means of discrete states.

The proposed algorithm is suitable for the cat-mouse-example. The design of controllers for real technical systems is much more complex. As an example, control engineers, who automate chemical plants, have to incorporate structured knowledge about recipes, standard specifications, etc. [21]. Algorithms like the cat-mouse-algorithm only cover specific problems of a design process.

4. THE ALGORITHM

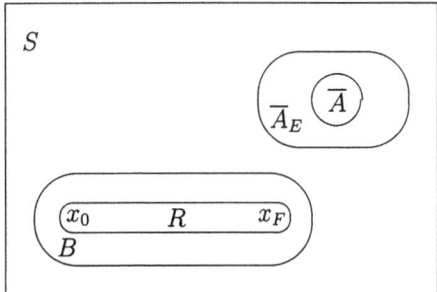

Figure 4: Venn diagram of the calculated state spaces.

1. Calculate extended forbidden state space $\overline{A}_E$.

2. Calculate coreachable state subset B of $S\backslash\overline{A}_E$, where S denotes the overall state space (backward calculation from final state x_F).

3. Calculate reachable state subset R of B (forward calculation from initial state x_0).

4. Calculate controller equations by means of the BDC.

Step 1: Assign the forbidden state space $\overline{A}$ (table 8) to

$$\overline{A}_E^0 := \overline{A}. \tag{27}$$

Calculate the extended forbidden state space $\overline{A}_E$ recursively with

$$\overline{A}_E^{k+1} := \max_{\mathbf{y},^1\mathbf{x},\mathbf{u}}^{l+n+m}(\Omega_c \cap \Omega_m$$
$$\cap \overline{A}_E^k \cap U_{uc}), \tag{28}$$

and stop if

$$\emptyset = \overline{A}_E^{k+1} \backslash \overline{A}_E^k, \tag{29}$$

where the difference $A\backslash B$ is the set of members of A which are not in B.

The set U_{uc} of uncontrollable events only contains c_7. Consequently, the controllable events $m_1 - m_6$ and $c_1 - c_6$ do not occur.

In almost all cases a relation that holds for

$$0 = f(\mathbf{x}) \tag{30}$$

is also true for

$$0 = f(^1\mathbf{x}). \tag{31}$$

Therefore the extended forbidden state space $\overline{A}_E$ has to be formulated as a function of the current state $\mathbf{x}$ (see table 9) as well as the next state $^1\mathbf{x}$.

$^1x_{c1}$	$^1x_{c2}$	$^1x_{c3}$	$^1x_{m1}$	$^1x_{m2}$	$^1x_{m3}$
0	0	0	0	0	0
1	0	0	1	0	0
0	1	0	0	1	0
1	1	0	1	1	0
0	0	1	0	0	1

Table 8: Forbidden state space $\overline{A}$.

x_{c1}	x_{c2}	x_{c3}	x_{m1}	x_{m2}	x_{m3}
0	0	0	0	0	0
0	1	0	0	1	0
0	0	1	0	0	1
1	-	0	1	-	0

Table 9: Extended forbidden state space $\overline{A}_E$.

Step 2: Assign the final state X_F (table 10) to

$$B^0 := X_F. \tag{32}$$

Compute by means of the backward calculation rule

$^1x_{c1}$	$^1x_{c2}$	$^1x_{c3}$	$^1x_{m1}$	$^1x_{m2}$	$^1x_{m3}$
0	1	0	0	0	1

Table 10: Final state X_F.

$$B^{k+1} := \max_{y,^1x,u}{}^{l+n+m}(\Omega_c \cap \Omega_m$$
$$\cap A_E \cap U_{res} \cap B^k) \tag{33}$$

all states from which the final state X_F can be reached (table 11), and stop if

$$\emptyset = B^{k+1} \backslash B^k. \tag{34}$$

The set U_{res} expresses the constraint that only one event may occur at one instant.

x_{c1}	x_{c2}	x_{c3}	x_{m1}	x_{m2}	x_{m3}
1	-	0	0	0	-
-	1	0	0	0	-
-	-	0	0	0	1
0	1	0	1	-	0

Table 11: Coreach B.

Step 3: Assign the initial state X_0 (table 12) to

$$R^0 := X_0. \tag{35}$$

x_{c1}	x_{c2}	x_{c3}	x_{m1}	x_{m2}	x_{m3}
0	1	0	0	0	1

Table 12: Initial state X_0.

Calculate the state space R (table 13), which contains all 'good' states, recursively with

$$R^{k+1} := \max_{y,x,u}{}^{l+n+m}(\Omega_c \cap \Omega_m$$
$$\cap B \cap U_{res} \cap R^k) \tag{36}$$

and stop if

$$\emptyset = R^{k+1} \backslash R^k. \tag{37}$$

x_{c1}	x_{c2}	x_{c3}	x_{m1}	x_{m2}	x_{m3}
0	1	0	1	1	0
0	1	0	0	0	-
-	-	0	0	0	1

Table 13: State space R.

Step 4: Intersecting all transitions from R to $\overline{R}$ with the automata of the cat and the mouse

$$G_{con} := \max_{\mathbf{y},^l\mathbf{x}}{}^{l+n}(\Omega_c \cap \Omega_m \cap U_{res} \cap R(\mathbf{x}) \cap \overline{R}({}^1\mathbf{x})), \qquad (38)$$

we find the implicit Boolean equation $0 = g_{con}(\mathbf{x}, \mathbf{u})$ with the solution set G_{con} of table 14 for all events that have to be disabled.

x_{c1}	x_{c2}	x_{c3}	x_{m1}	x_{m2}	x_{m3}	c_1	c_2	c_3	c_4	c_5	c_6	c_7	m_1	m_2	m_3	m_4	m_5	m_6
0	1	0	1	1	0	0	0	1	0	0	0	0	0	0	0	0	0	0
0	1	0	0	0	0	0	0	1	0	0	0	0	0	0	0	0	0	0
1	1	0	0	0	1	0	0	0	0	1	0	0	0	0	0	0	0	0
0	1	0	0	0	0	0	0	0	0	0	0	0	1	0	0	0	0	0
-	0	0	0	0	1	0	0	0	0	0	0	0	0	0	0	0	1	0
1	-	0	0	0	1	0	0	0	0	0	0	0	0	0	0	0	1	0

Table 14: G_{con} with all events that have to be disabled.

In order to obtain a simple and realizable discrete event controller $\mathbf{u} = \mathbf{g}'_{con}(\mathbf{x})$, the implicit Boolean equation $0 = g_{con}(\mathbf{x}, \mathbf{u})$ has to be solved for $\mathbf{u}$. This step is necessary, because process control systems, which are currently used in the industry, are not able to deal with implicit boolean equations. They need explicit boolean equations that can easily be implemented.

As a consequence, a controller $\mathbf{g}'_{con}(\mathbf{x})$ for the events c_3, c_5, m_1 and m_5 (see table 14) is required. But before we can develop this controller, problems of observability and observer design should be discussed. Considering that these topics are beyond the scope of this paper, we assume that the positions of the cat and of the mouse can be measured with the output equations

$$y_c = \overline{x}_{c1}x_{c2}\overline{x}_{c3} \qquad (39)$$
$$y_m = \overline{x}_{m1}\overline{x}_{m2}x_{m3}. \qquad (40)$$

Therefore y_c (y_m) is equal to 1 if the cat (mouse) is in its initial room 2 (4). A thorough discussion of observability and observer design for on-line state estimation and failure recognition in the framework of the BDC will be published in [19].

Inserting the solution sets of the output equations (39) and (40) into table 14, we find the combinations of input and output signals of table 15 that have to be forbidden.

y_c	y_m	c_3	c_5	m_1	m_5
1	0	1	0	0	0
0	1	0	1	0	0
1	0	0	0	1	0
0	1	0	0	0	1

Table 15: Forbidden combinations of input and output signals.

In this particular example, we obtain the controller equations

$$y_c \bar{y}_m \rightarrow \bar{c}_3 \tag{41}$$
$$\bar{y}_c y_m \rightarrow \bar{c}_5 \tag{42}$$
$$y_c \bar{y}_m \rightarrow \overline{m}_1 \tag{43}$$
$$\bar{y}_c y_m \rightarrow \overline{m}_5 \tag{44}$$

as a possible solution for the cat-mouse-problem.

4.1   Numerical aspects

Working with solution sets of boolean functions has been proven to be a powerful numerical approach. Don't care elements significantly reduce space and time complexity of the XBOOLE algorithms. In addition, the inherent bit-level parallelism of digital computers is exploited to perform all operations very quickly.

On the lowest computational level every set T with 1,0 and don't care elements is represented with two lists A and B, whose elements are 1 and 0. Steinbach proved that the code of table 16 is the most efficient one [26].

T_{ij}	A_{ij}	B_{ij}
-	0	0
0	0	1
1	1	1

Table 16: Binary code of ternary sets.

As an example, the intersection of two ternary vectors

$$\mathbf{t} = \mathbf{t}_1 \cap \mathbf{t}_2 \tag{45}$$

is calculated with the vector operations (see figure 5)

$$\mathbf{a} = \mathbf{a}_1 \vee \mathbf{a}_2 \tag{46}$$
$$\mathbf{b} = \mathbf{b}_1 \vee \mathbf{b}_2 \tag{47}$$

if

$$\mathbf{0} = (\mathbf{a}_1 \oplus \mathbf{a}_2)\mathbf{b}_1\mathbf{b}_2, \tag{48}$$

otherwise

$$\mathbf{t} = [\ ]. \tag{49}$$

$t_1 = [-\ 1\ 0\ -]$ $t_2 = [1\ -\ 0\ -]$

$\downarrow$ $\downarrow$

$a_1 = [0\ 1\ 0\ 0]$ $a_2 = [1\ 0\ 0\ 0]$
$b_1 = [0\ 1\ 1\ 0]$ $b_2 = [1\ 0\ 1\ 0]$

$a = [1\ 1\ 0\ 0]$
$b = [1\ 1\ 1\ 0]$

$\downarrow$

$t = [1\ 1\ 0\ -]$

Figure 5: Intersection of two ternary vectors.

The time complexity of the intersection of two sets $T_1 \cap T_2$ is of order $O(n_{T_1} n_{T_2})$, where n_{T_1}, n_{T_2} are equal to the number of rows of T_1 and T_2. The influence of the number of columns on the time complexity can be approximated with a constant.

XBOOLE is able to perform most operations with sets that have many thousand rows, even on a PC. As a result, very large state spaces - the size of a state space depends on the number of rows and the number of dashes - can be handled.

Current research is devoted to the development of efficient algorithms for parallel computers [27]. An outcome of this work will be a more powerful version of XBOOLE.

5.   CONCLUSION

A new approach for the synthesis of DES has been proposed. DES are represented by a standard model that consists of Boolean equations. The Boolean Lagrangian, which can easily be derived from the standard model, provides the basis for very efficient numerical algorithms. The computational complexity is substantially reduced by means of solution sets of Boolean functions and don't care elements.

The framework presented in this paper has been successfully applied to real process control problems. In most technical systems discrete states are either measurable or easily accessible (e.g. states that are represented in a computer). Consequently, keeping the state space information has been found to be particularly useful. The Boolean Differential Calculus as a state space approach provides a powerful basis for the analysis and synthesis of DES.

REFERENCES

[1] Y. C. Ho. Dynamics of discrete event systems. *Proceedings of the IEEE*, 77(1):3–6, January 1989.

[2] R. Scheuring and H. Wehlan. Der Boolesche Differentialkalkül - Eine Methode zur Analyse und Synthese von Petri-Netzen. *at - Automatisierungstechnik*, 39:226–233, July 1991.

[3] T. L. Booth. *Sequential Machines and Automata Theory*. John Wiley, New York, 1967.

[4] P. E. Caines, R. Greiner, and S. Wang. Dynamical Logic Observers for Finite Automata. In *Proceedings of the 27th IEEE Conference on Decision and Control*, pages 226–233, Austin, Texas, December 1988.

[5] P. E. Caines and S. Wang. Classical and Logic based Regulator Design and its Complexity for Partially Observed Automata. In *Proceedings of the 28th IEEE Conference on Decision and Control*, pages 132–137, Tampa, Florida, December 1989.

[6] L.E. Holloway and B.H. Krogh. Synthesis of feedback control logic for a class of controlled Petri nets. *IEEE Transactions on Automatic Control*, 35(5):514–523, May 1990.

[7] P.J. Ramadge and W.M. Wonham. Supervisory control of a class of discrete event systems. *Siam J. Control and Optimization*, 25(1):206–230, Jan. 1987.

[8] Randy Cieslak, C. Desclaux, Ayman S. Fawaz, and Pravin Varaiya. Supervisory control of discrete-event processes with partial observations. *IEEE Transactions on Automatic Control*, 33(3):249–260, March 1988.

[9] R.D. Brandt, V. Garg, P. Kumar, F. Lin, S.I. Marcus, and W.M. Wonham. Formulas for Calculating Supremal Controllable and Normal Sublanguages. *Systems & Control Letters*, 15:111–117, 1990.

[10] C.H. Golaszewski and P.J. Ramadge. Boolean coordination problems for product discrete event systems. In *Proceedings of the 29th IEEE Conference on Decision and Control*, pages 3428–3433, Honolulu, Hawaii, December 1990.

[11] Y. Brave and M. Heymann. Stabilization of discrete-event processes. *Int. J. Control*, 51(5):1101–1117, 1990.

[12] S. Balemi. *Control of Discrete Event Systems: Theory and Application*. PhD thesis, Swiss Federal Institute of Technology (ETH), Zurich, May, 1992.

[13] S. B. Akers. On a theory of Boolean functions. *SIAM J.*, 7:487–498, 1959.

[14] A. Talantsev. On the analysis and synthesis of certain electrical circuits by means of special logical operators. *Automat. i telemeh.*, 20:898–907, 1959.

[15] D. Bochmann and C. Posthoff. *Binäre dynamische Systeme*. R. Oldenbourg Verlag, München, Wien, 1981.

[16] A. Thayse. *Boolean Calculus of Differences*, volume 101 of *Lecture Notes in Computer Science*. Springer Verlag, Berlin, New-York, 1981.

[17] D. Bochmann and B. Steinbach. *Logikentwurf mit XBOOLE*. Verlag Technik, Berlin, 1991.

[18] R. Scheuring, B. Steinbach, and D. Bochmann. XBMini - User's Guide. Institut für Systemdynamik und Regelungstechnik, Stuttgart University, Internal report, 1992.

[19] R. Scheuring. *Modellierung, Beobachtung und Steuerung ereignisorientierter verfahrenstechnischer Systeme*. PhD thesis, Institut für Systemdynamik und Regelungstechnik, Stuttgart University, In preparation, 1993.

[20] P.J. Ramadge and W.M. Wonham. On the supremal controllable sublanguage of a given language. *SIAM J. Control and Optimization*, 25(3):637–659, May 1987.

[21] R. Scheuring and H. Wehlan. On the design of discrete event dynamic systems by means of the boolean differential calculus. In *First IFAC Symposium on Design Methods of Control Systems*, pages 723–728, Zurich, September 1991.

[22] W.V. Quine. A way to simplify truth functions. *American Mathematical Monthly*, 62(9):627–631, November 1955.

[23] E.J. McCluskey Jr. Minimization of boolean functions. *Bell System Technical Journal*, 35(6):1417–1444, November 1956.

[24] P. Kozák, S. Balemi, and R. Smedinga, editors. *Preprints of the Joint Workshop on Discrete Event Systems*, Prague, August 1992.

[25] J. L. Peterson. *Petri Net Theory and the Modeling of Systems*. Prentice-Hall Inc., Englewood Cliffs, New Jersey, 1981.

[26] B. Steinbach. Theorie, Algorithmen und Programme für den rechnergestützten logischen Entwurf digitaler Systeme. Dissertation B. Technische Universität Karl-Marx-Stadt, 1984.

[27] B. Steinbach and N. Kuemmling. Effiziente Lösung hochdimensionaler BOOLEscher Probleme mittels XBOOLE auf Transputer. In Grebe R. and C. Ziemann, editors, *Parallele Datenverarbeitung mit dem Transputer. Proceedings X (Springerreihe Informatikberichte Band 272)*, pages 127–134. Springer-Verlag, Berlin, New York, 1991.

A UNIFYING FRAMEWORK FOR DISCRETE EVENT SYSTEM CONTROL THEORY

Petr Kozák*

Abstract. An introduction to a unifying framework for discrete event systems control theory is presented. The framework is suitable for formulation and solution of control problems related to time discrete event systems at the qualitative level. Basic underlying ideas, definitions, and results are briefly presented. Several illustrative examples are provided.

1. INTRODUCTION

A unifying framework for discrete event systems is very desired, as there are many different approaches (see [1, 2, 3, 4] for a partial list) which, however, do not fully satisfy requirements of applications. Besides complexity issues, the main shortage of the present approaches is the highly idealised assumption of time-dependent system behaviour. The paper presents an introduction to a unifying DES theory based on the consistent utilization of the idea of input/output system modelling introduced within the system theory [5, 6, 7, 8]. Before we formulate the requirements of the framework, we shall classify shortly the classes of systems.

The models of discrete event systems (DES) (and also of general systems) can be classified according to the way of modelling system non-determinism into two groups:

- qualitative models,
- quantitative models.

The qualitative models give only a list of possibilities (e.g. finite automata) while the quantitative ones introduce a measure of non-determinism (e.g. probability in queuing networks).

Another important classification of DES models is according to the way of modelling the time. The following two basic classes can be recognised:

- logical models,
- time (temporal, real-time, ...) models.

The logical models describe only time order of events, e.g. automata, Petri nets. The time models provide in addition the time moments when the discrete events are observed, e.g. timed Petri nets.

*Czechoslovak Academy of Sciences, Institute of Information Theory and Automation, Pod vodárenskou věží 4, 182 08 Prague 8, Czechoslovakia; e-mail: kozak@cspgas11.bitnet
The work was partially supported by the grants ČSAV #27502 and ČSAV #27558, and Bell Canada research contract No. 3-254-188-10 (Univ. of Toronto). The paper was finished while the author was on leave at the Systems Control Group, Dept. of Electrical Engineering, University of Toronto, Toronto, Ontario, M5S 1A4 Canada; e-mail: kozak@odin.control.utoronto.ca

The aim of the paper is to present basic ideas underlying to the unifying framework of time DES control at the qualitative level [9, 10, 11, 12, 13, 14]. Any discrete event system can be modelled using techniques introduced for simulation of systems. But these models are not usually suitable for mathematical analysis, i.e. also not for synthesis of controllers. The new unifying framework aims not only at modelling general discrete event systems, but mainly at providing a methodology for proving results at a level as general as possible. In this way, DES theory can be made more effective and its applicability can be enlarged.

A suitable unifying framework should satisfy the following requirements:

1. The concepts of the framework should be compatible with the traditional system theory.

2. The basic definitions (e.g. of system, feedback interconnection, etc.) should be so general that they can be used also for continuous variable systems (CVS) (which are usually described by differential or difference equations). It would enable to develop a framework suitable also for "hybrid" systems incorporating features of both DESs and CVSs.

3. Simple transformations between time and logical DES models should be possible.

4. Possibility of extension to the quantitative level.

5. The framework should provide a general technique for effective design of feedback controllers.

6. Suitability for development of modular and/or hierarchical models and methods within the framework.

The paper presents an attempt to develop a framework satisfying the above given requirements [9, 10, 11, 12, 13, 14].

Sections 2. and 3. introduce a model of time systems and some necessary notation. Section 4. gives definition of controllable behaviour and feedback interconnection. The concept of control and output laws is defined in section 5. Certain basic results are presented and discussed. Section 6. formulates a sample control problem and outlines the utilization of the presented framework in controller design. Section 7. presents some results necessary for design of strongly causal controllers within the framework. The solution of the cat-and-mouse problem [15] is suggested as a concrete non-trivial simple example. The example reflects certain situations arising in manufacturing systems. The animals can be replaced e.g. by automatically guided vehicles.

2. TIME SYSTEMS

Let us consider a manufacturing system consisting of several NC machines, manipulators, automatically guided vehicles (AGV), and of other special equipment. Several levels of modelling and of control problems can be recognised, starting from on-board sensing and control of particular machines up to the supervisory level of control. In the following, we shall focus on the supervisory level, but applications of DES methods and models are not limited only to this level.

The supervisor is connected with the manufacturing system via a communication network transmitting packets with information that can be interpreted using the following sentences: "The NC-machine x has started operation y.", "The AGV x stopped in the position y.", etc. The supervisor applies commands like "Let no AGV enter the area x.", "Let the heater x start up.", etc. These sentences are represented by labels of discrete events.

One of the important parameters of the system is the set of discrete event labels, which can be divided into the set of output discrete event labels

$$Y \ = \ \{ \ \text{"The NC-machine } x \text{ have started operation } y \text{.",}$$
$$\text{"The AGV } x \text{ stopped at the position } y \text{." } \}$$

and the set of the input discrete event labels

$$U \ = \ \{ \ \text{"Let no AGV enter the area } x \text{.",}$$
$$\text{"Let the heater } x \text{ start up." } \}$$

We denote the set of discrete event labels as

$$\Sigma = \{e_i\}_{i \in I} \ \ [= U \cup Y],$$

where I is an index set (not necessarily finite) and e_i $(i \in I)$ are event labels.

A very natural model of a manufacturing system is the set of all sequences of discrete event labels (written in the order of occurrence), which can be observed on the system. These sequences are called logical observations. As simultaneous occurrence of several discrete events at the same moment is possible, it is convenient to model the system as a set of sequences over the power set of Σ, i.e. the logical behaviour of the system is

$$\mathcal{G}_L = \{w_j\}_{j \in J},$$

where J is an index set and for all $j \in J$ the elements of w_j are in $2^\Sigma \setminus \{\emptyset\}$. The power set of Σ is denoted by 2^Σ and the empty set by $\emptyset$.

For instance the sequence (for some $j \in J$)

$$w_j = \{e_3\}\{e_2, e_{10}\}\{e_6\}$$

is interpreted such that at first an event labelled by e_3 is observed, then a simultaneous event consisting of two events labelled e_2 and e_{10} occurs, and then an event labelled by e_6 occurs.

The above presented model is in the class of logical DES models. Supposing that Σ is finite then $\mathcal{G}_L$ is a formal language that can be represented as an automaton, or Petri net, etc. If conditional probabilities of event occurrences are added a Markov chain can be obtained.

Let us remark that the above given explanation of DES models is simplified a little but it is sufficient for our purpose. A time model of the given DES can be obtained by providing time moments for each element of each sequence of the logical behaviour such that the time moments of each sequence are increasing (this is necessary in order to be consistent with the logical model).

Supposing that we are modelling the manufacturing system over a linearly ordered time set T (equal usually to non-negative reals $T = \mathbb{R}_0^+$, but it can be equal to the set of natural numbers $\mathbb{N}$ as well), the time behaviour of the system can be modelled as

$$\mathcal{G} = \{o_k : T \mapsto 2^\Sigma\}_{k \in K}, \qquad (1)$$

where K is an index set and for all $o \in \mathcal{G}$

$$\text{the set } \{t \in T : o(t) \neq \emptyset\} \text{ is at most countable.} \qquad (2)$$

In this model the empty set plays the role of "null (empty) event" modelling the signal observed on the input and the output if no discrete event occurs. Let us note that K is in general a much greater set than J, because to each logical observation, there can be several (even infinite number of) corresponding time observations.

Up to now, we have been using the term DES without any appropriate definition. The most important feature characterising DESs is the condition (2), which can be interpreted roughly in such a way that the discrete events can be observed in at most countable time instances. The continuous variable systems (CVS) (modelled usually by means of differential equations) do not satisfy this condition in general.

Another important feature of DESs is the non-numerical nature of Σ. It means roughly that even if $\Sigma \subseteq \mathbb{R}$ ($\mathbb{R}$ denotes the set of real numbers) then no interpretation of addition and multiplication of event labels can be found. Therefore, the models utilized in discrete time control cannot be usually used. Nevertheless, some important and interesting analogies can be found (e.g. with linear systems [16]).

Note that the time models of DESs are usually constructed on the basis of logical models by providing some mappings describing time delays (e.g. minimal and maximal delays between two subsequent output discrete events). These models, however, do not satisfy the requirements given in section 1. We shall continue in the direction given by model (1).

Model (1) can be expressed more compactly as (let us call it a time system – TS.) an ordered quintuple

$$\mathcal{S} = (T, \leq, U, Y, \mathcal{G}),$$

where T, U and Y are sets, $\leq$ is a linear order in T, $U \cap Y = \emptyset$, and $\emptyset \neq \mathcal{G} \subseteq \mathcal{F}(T, 2^{U \cup Y})$. The symbols $\mathcal{F}(V, W)$ denote the set of all mappings from V to W defined everywhere. If, in addition, $\mathcal{G}$ satisfies condition (2) for all $o \in \mathcal{G}$ then $\mathcal{S}$ is called time DES (TDES).

The ordered set $(T, \leq)$ plays the role of a time set. The set $\mathcal{G}$ is the (time) behaviour of $\mathcal{S}$. Each mapping $o \in \mathcal{G}$ is interpreted as an input/output (time) observation. For all $t \in T$ the set $o(t)$ models the value observed on the input and the output during the observation o at the time moment t. Note that the set $o(t)$ can model simultaneous discrete events (if $\text{card}(o(t)) > 1$, where card means the cardinality of a set) or "null (empty)" event (if $o(t) = \emptyset$). Here we suppose that the input and output discrete event labels are disjoint. A more general formulation of time system definition and a general model of simultaneous events are presented in [9]. The definition of system given above is the same in nature as the definitions [5, 6, 7, 8]. The difference is in its utilization and consequently almost all other definitions starting from the feedback interconnection have to be different (section 4.).

Note that we are not using the concept of a state as the primary concept in the definition of TS (as usually done). This way of modelling a system is called input/output modelling. It requires less notation at this general level, it enables more compact formulas and moreover it is proved to have the same expressive power [9]. If required, the concept of a state can be introduced as a secondary one [9]. Certain high-level constructs for specifying system behaviours are introduced in section 5.

The transformation mapping between the time observations in $\mathcal{G}$ and the logical observations is defined in a natural way [9, 12]. Let $\mathcal{S} = (T, \leq, U, Y, \mathcal{G})$ be a TDES. We define for all $f \in \mathcal{G}$ an increasing sequence $\tilde{s}(f)$ such that $\text{Im}(\tilde{s}(f)) = \{t \in T : f(t) \neq \emptyset\}$. The symbol "Im" denote the image of a mapping. Note that the image of a sequence is a union of its elements. The sequence $\tilde{s}(f)$ represents time moments of discrete event occurrences. The behaviour $\mathcal{G}$ can be transformed into a corresponding logical behaviour using a mapping defined as follows. For all $f \in \mathcal{G}$ we define a sequence $s(f)$ such that $\text{Dom}(s(f)) = \text{Dom}(\tilde{s}(f))$ and for all $i \in \text{Dom}(s(f))$ we have $s(f)(i) = f(\tilde{s}(f)(i))$. The elements $s(f)(i)$ and $\tilde{s}(f)(i)$ stand here for the i-th member of the sequence $s(f)$ and $\tilde{s}(f)$, respectively. A logical DES corresponding to the TDES $\mathcal{S}$ can be defined using the logical behaviour $\mathcal{G}_L = s(\mathcal{G})$.

To play with the sets of behaviours, the following simple notation is given in [9, 10, 11, 12, 13, 14]. It is also required for the definition of feedback interconnection and of output and control laws.

During the observation of a TS $\mathcal{S} = (T, \leq, U, Y, \mathcal{G})$ we are observing on the input and output just only trajectories corresponding to the initial parts of the mappings in its behaviour. These parts form a set of all prefixes of $\mathcal{G}$. The domains of these prefixes are initial segments of T. Formally: a set $A \subseteq T$ is called an *initial segment of* $(T, \leq)$ if for all $a \in A$ and $t \in T$ such that $t \leq a$ we have $t \in A$. The set of all initial segments of $(T, \leq)$ is denoted by $\mathcal{I}_{(T, \leq)}$. Let $\mathcal{M} \subseteq \mathcal{G}$. We define the set of all prefixes of $\mathcal{M}$ as

$$\text{pref}(\mathcal{M}) = \{f \,|\, A : \ f \in \mathcal{M} \ \wedge \ A \in \mathcal{I}_{(T, \leq)} \}.$$

An arbitrary initial segment A is *upper bounded* if there exists $t \in T$ such that $a \leq t$ for all $a \in A$. The set of all upper bounded initial segments of $(T, \leq)$ is denoted by $\mathcal{I}^b_{(T, \leq)}$.

For an arbitrary mapping o we define

$$\mathcal{O}_{\mathcal{S}}(o) = \{f \in \mathcal{G} : \ f \,|\, \text{Dom}(o) = o \},$$

where $\text{Dom}(o)$ is the domain of o and $|$ denotes the restriction of mappings. The restriction $f \,|\, \text{Dom}(o)$ is defined as a mapping with the domain equal to $\text{Dom}(o) \cap \text{Dom}(f)$.

Supposing that $o \in \text{pref}(\mathcal{G})$, the set $\mathcal{O}_{\mathcal{S}}(o)$ consists of all observations in $\mathcal{G}$ which can be observed on $\mathcal{S}$ if o has been observed. So the set of all possible future trajectories is given in this way. It can be extended for sets in a natural way. Let $\mathcal{R} = \{r_m\}_{m \in M}$ is a set of mappings with an index set M. We define

$$\mathcal{O}_S(\mathcal{R}) = \{f \in \mathcal{G} : \exists m \in M, \ f \,|\, \mathrm{Dom}(r_m) = r_m \ \}.$$

Each observation $o \in \mathcal{G}$ contains information about both input and output observations (trajectories). The following mapping is suitable for abstraction of information from the observations. It is in fact a masking operation. For an arbitrary set V we define $m_V \circ \mathcal{M} = \{m_V \circ f : f \in \mathcal{M}\}$, where $m_V \circ f : T \mapsto 2^{U \cup Y}$ such that for all $t \in T$ it holds $m_V \circ f(t) = f(t) \cap V$. The circle "$\circ$" denotes composition of mappings, i.e. m_V is defined as a mapping which maps each set into its intersection with V.

For an arbitrary set V we define

$$\mathcal{O}_S^V(\mathcal{R}) = \{f \in \mathcal{G} : \exists m \in M, \ m_V \circ f \,|\, \mathrm{Dom}(r_m) = r_m \ \}.$$

For each $u \in \mathrm{pref}(m_U \circ \mathcal{G})$ (i.e. for each initial part of each input observation) the set $\mathcal{O}_S^U(\{u\})$ represents the set of all mappings having the initial part of input observation equal to u. We write usually $\mathcal{O}_S^U(u)$ instead of $\mathcal{O}_S^U(\{u\})$.

It is often convenient to restrict attention to certain subclasses of TDESs. A TDES $\mathcal{S} = (T, \leq, U, Y, \mathcal{G})$ is ordinary if it satisfies for all $o \in \mathcal{G}$ the condition

$$\forall A \in \mathcal{I}_T^b, \text{ the set } \{t \in A : o(t) \neq \emptyset\} \text{ is finite.} \tag{3}$$

It is finitary ordinary TDES if for all $o \in \mathcal{G}$

$$\text{the set } \{t \in T : o(t) \neq \emptyset\} \text{ is finite,} \tag{4}$$

i.e. in each time interval only a finite number of discrete events can be observed.

It is possible in general that a TS $\mathcal{S} = (T, \leq, U, Y, \mathcal{G})$ has for some $u \in m_U \circ \mathcal{G}$ several possible output trajectories corresponding to this input trajectory, i.e. $\mathrm{card}(\mathcal{O}_S^U(u)) > 1$. Such system is called non-deterministic.

To illustrate the concepts, let us consider a finitary ordinary TDES $\mathcal{S}_x = (T, \leq, U, Y, \mathcal{G}_x)$ defined as follows. Let $U = \{e, d\}$ and $Y = \{a, b\}$. Let L is a language given by the regular expression $(\{a\}\{b\})^*$, i.e. L consists of words $\{a\}\{b\}$, $\{a\}\{b\}\{a\}\{b\}$, etc. Let us remark that the event labels are sets in the model, therefore L is a language over the alphabet $\Gamma = \{\{a\}, \{b\}\}$. The language L represents the logical output behaviour of the system $(s(m_Y \circ \mathcal{G}_x))$.

Each event labelled by d disables all future occurrences of events labelled by a unless an input event labelled by e enabling them again is applied. It is supposed that all output events are enabled before the first input event is applied. The occurrences of events labelled by b cannot be controlled by means of input signal.

The system $\mathcal{S}_x$ is modelled over the time set of the non-negative real numbers with the canonical order. Let $\mathcal{A}$ be a set of all mappings o in $\mathcal{F}(T, 2^{U \cup Y})$ such that condition (4) is satisfied.

Let us suppose that no output discrete event can be generated at the time moment 10. No other special property of time delays is supposed.

The behaviour $\mathcal{G}_x$ can be defined as follows:

$$\mathcal{G}_x = \{ f \in \mathcal{A} : \; s(m_Y \circ f) \in L \; \wedge$$
$$\wedge \; (\; \forall t \in T, \; m_Y \circ f(t) = \{a\} \; \Rightarrow \; (10 < t \; \wedge \; d \notin m_U \circ f(z) \;)) \; \},$$

where $z = \max\{\tau \in T : \; \tau = 0 \; \vee \; (\tau < t \; \wedge \; m_U \circ f(\tau) \neq \emptyset) \; \}$.

Let $u_x \in m_U \circ \mathcal{G}$ such that $u_x(0) = \{d\}$ and for all $t \in T$, $t \neq 0$ it holds $u_x(t) = \emptyset$. The mapping u_x represents the input trajectory consisting just only from one discrete event labelled by "d" at the time moment 0. It can be easily verified that $m_Y \circ \mathcal{O}^U_{\mathcal{S}_x}(u_x) = \{y\}$ for some y where for all $t \in T$ $y(t) = \emptyset$, i.e. no output discrete event is observed if the control u is applied. In other words $s(m_Y \circ \mathcal{O}^U_{\mathcal{S}_x}(u_x))$ is equal to the empty word. But $s(\mathcal{O}^U_{\mathcal{S}_x}(u_x))$ is equal to the word $\{d\}$. It can be also verified that $\mathcal{S}_x$ is non-deterministic, because *e.g.* there is an infinite set of output trajectories corresponding to the empty input trajectory.

3. SOME NATURAL SYSTEM PROPERTIES

Usually we suppose that real-world systems have certain properties which are believed to be natural. One of such properties is causality. A system is "causal" if it generates output trajectories as a response to the previously observed input trajectory (and of course to previous output trajectory), i.e. independently of the future input trajectory. As "non-causal" systems can be also easily modelled using the formalism given above, it is necessary to define formally the concept of causality.

A TS $\mathcal{S} = (T, \leq, U, Y, \mathcal{G})$ is causal if for all $u, u' \in m_U \circ \mathcal{G}$ and $A \in \mathcal{I}^b_T$

$$u \,|\, A = u' \,|\, A \; \Rightarrow \; \mathcal{O}^U_{\mathcal{S}}(u) \,|\, A = \mathcal{O}^U_{\mathcal{S}}(u') \,|\, A,$$

i.e. for all input trajectories u, u' that are equal on A the sets of possible corresponding output trajectories restricted to A are the same. A causal system is able, in general, to give a response to a value observed on its input at the same time moment when this input value is observed. It is highly unrealistic. Therefore, the concept of strong causality is introduced.

Let us denote for all $A \in \mathcal{I}^b_T$ such that $\sup_{(T, \leq)}(A)$ exists the set $\tilde{A} = A \cup \sup_{(T, \leq)}(A)$, where $\sup_{(T, \leq)}(A)$ is the least upper bound of A with respect to $(T, \leq)$.

The TS $\mathcal{S}$ is strongly causal if it is causal and for all $u, u' \in m_U \circ \mathcal{G}$ and $A \in \mathcal{I}^b_T$ such that $\sup_{(T, \leq)}(A)$ exists the following holds

$$u \,|\, A = u' \,|\, A \; \Rightarrow \; m_Y \circ \mathcal{O}^U_{\mathcal{S}}(u) \,|\, \tilde{A} = m_Y \circ \mathcal{O}^U_{\mathcal{S}}(u') \,|\, \tilde{A}.$$

Definitions of causality and strong causality can be found *e.g.* in [5], where they are given using the concept of a state. The definitions presented here and in [9, 12, 14] are formulated using consistent input/output modelling of systems. Certain basic results related to these two concepts can be found in [9].

It can be verified that the TDES $\mathcal{S}_x$ of the previous section is strongly causal.

In the two following sections, we shall focus on the design of causal controllers. Analogous results for strongly causal controllers are in section 7.

4. FEEDBACK INTERCONNECTION AND CONTROLLABLE BEHAVIOURS

Another concept which is necessary for any control theory is that of a feedback interconnection of systems. The proposed controller design method uses the concepts of output and control laws. The feedback interconnection is necessary for proving the validity of this procedure. However, the knowledge of the definition is not necessary for applications of the procedure. Therefore, a reader who is interested primarily in the application of the proposed design procedure can jump to the last paragraph of this section.

Unfortunately, the only one definition of feedback interconnection given using an input/output approach [5] is proved to be inconsistent with intuition [9] (e.g. for certain non-deterministic systems modelled at the qualitative level), therefore it cannot be used in controller design. The definition in [5] can be formulated in our notation as follows. Let $S_1 = (T, \leq, U, Y, \mathcal{G}_1)$ and $S_2 = (T, \leq, Y, U, \mathcal{G}_2)$ be TSs. The behaviour of feedback interconnection of S_1 and S_2 is given by $\mathcal{G} = \mathcal{G}_1 \cap \mathcal{G}_2$. According to this definition, arbitrary systems can be interconnected. Moreover it is not necessary that the systems generate their outputs according to the signal observed previously on their inputs. The output signal can depend also on a future input signal. To avoid these difficulties three new additional axioms are formulated in [9]. To simplify the definition, we shall introduce at first the concept of controllable behaviour. It is shown in [9] that the behaviour of a TS S can be reduced to any controllable behaviour $\mathcal{G}'$ of S by means of a suitable causal feedback controller.

Let us consider an arbitrary TS $S = (T, \leq, U, Y, \mathcal{G})$, $\mathcal{G}' \subseteq \mathcal{G}$ and the system $S' = (T, \leq, U, Y, \mathcal{G}')$. The set $\mathcal{G}'$ is a *controllable behaviour of S* if the following two conditions are satisfied:

Free Output Continuation Choice Axiom. For all $f \in \mathrm{pref}(\mathcal{G}')$ and $y \in m_Y \circ \mathcal{O}_S(f)$ the following condition

$$\exists A \in \mathcal{I}_T, \ \mathrm{Dom}(f) \subset A \ \wedge \ y \,|\, A \in m_Y \circ \mathcal{O}_S^U(m_U \circ \mathcal{O}_{S'}(f)) \,|\, A,$$

implies that

$$\exists B \in \mathcal{I}_T, \ \mathrm{Dom}(f) \subset B \ \wedge \ y \,|\, B \in m_Y \circ \mathcal{O}_{S'}(f) \,|\, B.$$

It is interpreted such that it must always (i.e. for all $f \in \mathrm{pref}(\mathcal{G}')$) hold that any possible future output observation (i.e. y) of S which is enabled at least by one possible future input of S' even for some very short future time must be contained in the possible future output observations of S' at least for some short future time interval.

Strict End Existence Axiom. For all $f \in \mathrm{pref}(\mathcal{G}')$, $y \in m_Y \circ \mathcal{O}_S(f)$ and $A \in \mathcal{I}_T$ such that $\mathrm{Dom}(f) \subset A$ and $y \,|\, A \notin m_Y \circ \mathcal{O}_{S'}(f) \,|\, A$ it holds that

$$\exists g \in \mathrm{pref}(\mathcal{G}'), \ f \in \mathrm{pref}(g) \ \wedge \ m_Y \circ g = y \,|\, \mathrm{Dom}(g) \ \wedge \ \mathrm{Dom}(g) \subset A \ \wedge$$
$$(\forall B \in \mathcal{I}_T, \ \mathrm{Dom}(g) \subset B \Rightarrow y \,|\, B \notin m_Y \circ \mathcal{O}_S^U(m_U \circ \mathcal{O}_{S'}(g)) \,|\, B \,).$$

This condition means that it must be always satisfied (i.e. again for all $f \in \mathrm{pref}(\mathcal{G}')$) that if some possible future output observation (i.e. y) of S is not contained in the future output observations of S' then there must exist some "future" time interval (i.e. $\mathrm{Dom}(g)$) such that no future input observation of S' does enable y in S even for some short time interval.

A TS $S = (T, \leq, \emptyset, Y, \mathcal{G})$ is a feedback interconnection of $S_1 = (T, \leq, U, Y, \mathcal{G}_1)$ and $S_2 = (T, \leq, Y, U, \mathcal{G}_2)$ (figure 1) if the following conditions are satisfied

(i) **Common Prefix Preservation Axiom**
$$\mathrm{pref}(\mathcal{G}_1) \cap \mathrm{pref}(\mathcal{G}_2) = \mathrm{pref}(\mathcal{G}_1 \cap \mathcal{G}_2),$$

(ii) $\mathcal{G}_1 \cap \mathcal{G}_2$ is a controllable behaviour of S_1,

(iii) $\mathcal{G}_1 \cap \mathcal{G}_2$ is a controllable behaviour of S_2,

(iv) $\mathcal{G} = m_Y \circ (\mathcal{G}_1 \cap \mathcal{G}_2)$.

It can be verified that the TDES S_x of section 2. can be interconnected in the feedback with a system $S'_x = (T, \leq, Y, U, \mathcal{G}'_x)$, where $m_U \circ \mathcal{G}'_x = \{u_x\}$ (u_x is defined in the same section) and $m_Y \circ \mathcal{O}^U_{S'_x}(u_x) = \mathcal{F}(T, 2^Y)$. The behaviour of the coupled system consists only of one observation with no output discrete event.

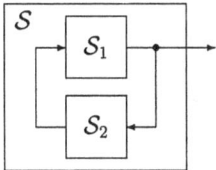

Figure 1:

A typical control problem solved in control theory can be formulated as follows. Find a feedback controller S_2 (figure 1) within a given class of controllers such that the behaviour of feedback interconnection of a given TS S_1 and of S_2 satisfies given constraints. The necessary and sufficient condition for existence of a causal controller which realises a behaviour $\mathcal{G}'$ if interconnected with S_1 is that $\mathcal{G}'$ is a controllable behaviour of S_1. This is proved in a constructive way in [10]. Therefore, each control problem can be reformulated as a problem of determining a controllable behaviour of the given system which satisfies the given constraints. Moreover, it is shown (again in a constructive way), that if S_1 is a TDES (ordinary or finitary ordinary TDES) then the controller can be also a TDES (ordinary or finitary ordinary TDES, respectively).

5. HIGH-LEVEL CONSTRUCTS FOR SYSTEM MODELLING

The results concerning the existence of controllers mentioned in the previous section can be used for controller design. However, proving system properties is not that much effective and also constructing sets describing the system behaviour is ineffective. Therefore, high-level constructs called output and control laws are introduced in [11].

The main idea of output and control laws is to model systems using a set of observations $\mathcal{M}$ and a mapping (called law) from $\text{pref}(\mathcal{M})$ into the power set of the output or input discrete event labels, respectively. The behaviour $\mathcal{G}$ of the time system $\mathcal{S}$ is formed by all observations in $\mathcal{M}$ such that the output values (or input values, respectively) are given by the mapping (*i.e.* output or control law, respectively). This technique is called semi-state [11] as the form of output and control laws is similar to the state space form. However, it does not contain all information about the system as it is in the state space modelling. The additional information is provided by the set $\mathcal{M}$. The advantage is a relatively simple formulation of system behaviour with great expressive power and with the possibility of simple proving of general results.

Using this technique, sufficient conditions for existence of feedback controllers are proved in a constructive way for a large class of systems. This methodology gives an efficient basis for modelling and control of TSs.

Let us consider a time set $(T, \leq)$ and sets of discrete event labels U and Y. Let $\Sigma = U \cup Y$ and $\mathcal{M} \subseteq \mathcal{F}(T, 2^\Sigma)$ in this section. A law is mapping from $\mathcal{M} \times T$ to $2^{2^\Sigma} \setminus \{\emptyset\}$, where $\setminus$ is the set difference. For the definition of input and output laws we need the following notation. For all $t \in T$

$$\begin{aligned}
\mathcal{J}_t &= \{\tau \in T : \tau \leq t\} \\
\mathcal{J}_t^s &= \{\tau \in T : \tau \leq t \wedge \tau \neq t\}.
\end{aligned}$$

5.1 Output Laws

A law $\mu_o : \mathcal{M} \times T \mapsto 2^{2^Y} \setminus \{\emptyset\}$ is an *output law* if for all $f, g \in \mathcal{M}$ and $t \in T$ the following condition

$$m_U \circ f \,|\, \mathcal{J}_t = m_U \circ g \,|\, \mathcal{J}_t \quad \wedge \quad m_Y \circ f \,|\, \mathcal{J}_t^s = m_Y \circ g \,|\, \mathcal{J}_t^s$$

implies that $\mu_o(f, t) = \mu_o(g, t)$.

A TS $\mathcal{S} = (T, \leq, U, Y, \mathcal{G})$ can be defined using $\mathcal{M}$ and the output law μ_o such that

$$\mathcal{G} = \{f \in \mathcal{M} : \forall t \in T, \, m_Y \circ f(t) \in \mu_o(f, t) \}. \tag{5}$$

In this way, TDESs can be modelled using a set $\mathcal{M}$ also satisfying condition (2) for all $o \in \mathcal{M}$. Analogously, the ordinary and finitary ordinary TDESs can be modelled using $\mathcal{M}$ satisfying condition (3) respectively (4) for all $o \in \mathcal{M}$. The formulation of the "cat-and-mouse model" in [15] shows how the concept of output law can be used in modelling.

The output laws can be constructed in a modular way like it is shown in [12].

The following result proved in [11] gives a sufficient condition for several important technical system properties. If this sufficient condition is fulfilled by the model of the controlled system, then the verification of controllers can be much simplified as it is shown in [11].

A TS $\mathcal{S} = (T, \leq, U, Y, \mathcal{G})$ modelled using $\mathcal{M}$ and μ_o satisfies the following three conditions

(*i*) $\mathcal{S}$ is causal,

(*ii*) $m_U \circ \mathcal{G} = m_U \circ \mathcal{M}$,

(*iii*) $\forall f \in \mathrm{pref}(\mathcal{M})$,

$$(\forall t \in \mathrm{Dom}(f),\ m_Y \circ f(t) \in \mu_o(f,t)) \ \Rightarrow \ f \in \mathrm{pref}(\mathcal{G})$$

if it holds for all $f \in \mathrm{pref}(\mathcal{M})$

$$(\forall t \in \mathrm{Dom}(f),\ m_Y \circ f(t) \in \mu_o(f,t))$$
$$\Rightarrow (\forall u \in m_U \circ \mathcal{O}^U_{\mathcal{S}'}(m_U \circ f),\ \exists g \in \mathcal{O}_{\mathcal{S}}(f),\ m_U \circ g = u), \quad (6)$$

where $\mathcal{S}' = (T, \leq, U, Y, \mathcal{M})$.

Condition (6) looks complicated but it is very often trivially satisfied by many systems [12, 15]. It can be interpreted as follows. Supposing that f satisfies the output law, then for all possible future input trajectories in $\mathcal{M}$ there exists some future output response in $\mathcal{G}$.

5.2   Control Laws

A law $\mu_c : \mathcal{M} \times T \mapsto 2^{2^U} \setminus \{\emptyset\}$ is a *control law* if for all $f, g \in \mathcal{M}$ and $t \in T$ the following condition

$$m_U \circ f \,|\, \mathcal{J}^s_t = m_U \circ g \,|\, \mathcal{J}^s_t \ \wedge \ m_Y \circ f \,|\, \mathcal{J}_t = m_Y \circ g \,|\, \mathcal{J}_t$$

implies that $\mu_c(f,t) = \mu_c(g,t)$.

A behaviour $\mathcal{G}$ can be defined on the basis of $\mathcal{M}$ and μ_c analogously to the modelling of systems using output laws. A sufficient condition for testing if $\mathcal{G}$ is controllable behaviour of $\mathcal{S}' = (T, \leq, U, Y, \mathcal{M})$ will be presented.

We define

$$\mathcal{G} = \{ f \in \mathcal{M} : \ \forall t \in T,\ m_U \circ f(t) \in \mu_c(f,t) \}. \quad (7)$$

Let us consider an ordinary TDES $\mathcal{S}' = (T, \leq, U, Y, \mathcal{M})$ with its behaviour constructed on the basis of a set $\mathcal{M}'$ and an output behaviour μ'_o in the way presented in the previous subsection. Let us suppose that the system $\mathcal{S}'$ satisfies the sufficient condition (6) if we put μ'_o for μ_o, $\mathcal{M}'$ for $\mathcal{M}$ and $\mathcal{M}$ for $\mathcal{G}$. Note, that $\mathcal{M}'$ must satisfy condition (3), because we are modelling an ordinary TDES. The set $\mathcal{G}$ defined by (7) is a controllable behaviour of $\mathcal{S}'$ if

$$\forall f \in \mathrm{pref}(\mathcal{M}),\ (\forall t \in \mathrm{Dom}(f),\ m_U \circ f(t) \in \mu_c(f,t)) \ \Rightarrow \ f \in \mathrm{pref}(\mathcal{G}). \quad (8)$$

This result proved in [11] can be immediately used in controller design as it is suggested in the next section and in [12, 13, 15].

6.   A GENERAL SCHEME FOR CONTROLLER DESIGN

Let us formulate at first an example of a control problem, *e.g.* a time analogy of the logical DES control problem formulated in [17]. This time control problem is solved in [12] for a class of TDESs.

Let $\mathcal{S} = (T, \leq, U, Y, \mathcal{G})$ be a causal finitary ordinary TDES such that Y is a finite set. Let $L = s(m_Y \circ \mathcal{G})$. The language L represents the logical output behaviour of $\mathcal{S}$. Let $\emptyset \neq L_a \subseteq L_g \subseteq L_m \subseteq L$. These languages play here a role similar role to the languages in [17].

The control problem is to find a controllable behaviour $\mathcal{Z}$ of $\mathcal{S}$ such that the following two constraints are satisfied

$$L_a \subseteq s(m_Y \circ \mathcal{Z}) \cap L_m \subseteq L_g,$$
$$\forall f \in \mathcal{Z}, \forall A \in \mathcal{I}_T^b, \ \exists g \in \mathcal{O}_{\mathcal{S}'}(f), \ s(m_Y \circ g) \in L_m,$$

where $\mathcal{S}' = (T, \leq, U, Y, \mathcal{Z})$ is a TS.

The marked language L_m consists of marked sequences of discrete event labels, which can correspond to the tasks completed by the underlying real-world system. The first constraint requires that all marked sequences generated by $\mathcal{Z}$ are within certain bounds given by L_a (minimal acceptable output logical behaviour) and L_g (legal output logical behaviour).

The second constraint is analogous to the requirement of non-blockingness in [17], *i.e.* there is always the possibility of generating at least one marked sequence. It can be interpreted also as an absence-of-deadlock condition, supposing that the deadlock is defined as the impossibility to generate any marked sequence.

Let us suppose that we would like to solve the problem for the system $\mathcal{S}_x$ (section 2.) and that $L_a = L_g = L_m = \{\epsilon\}$, where ϵ is the empty word. It can be verified that one of the possible controllable behaviour solving the problem is $\mathcal{G}_x \cap \mathcal{G}'_x$, where $\mathcal{G}'_x$ is defined at the end of section 4.

Note that the output logical behaviour plays an important role in the constraints. But the controller must handle also time information, i.e. the behaviour $\mathcal{G}$. It is possible in general that for all $f \in \mathrm{pref}(\mathcal{G})$ the language $s(m_Y \circ \mathcal{O}_{\mathcal{S}}(f))$ is different from the language

$$\{w \in L : \ w = w'w'' \ \wedge \ w' = s(m_Y \circ f) \ \wedge \ w'' \in (2^Y)^* \},$$

where $(2^Y)^*$ is the set of all finite sequences over 2^Y. This fact is interpreted in such a way that the possible future output logical behaviour of a TDES is different in general from the future output behaviour of a corresponding logical DES.

The design of feedback controllers using the results mentioned above consists basically of five steps:

1. Specification of a general form of output laws describing the systems to be controlled and proof that this general form satisfies the sufficient conditions (section 5.). This form can have parameters like the logical behaviour of the uncontrolled system, a mapping describing least delays etc.

2. Formulation of a control problem.

3. Proof that a proposed control law satisfies the sufficient conditions for existence of controllable behaviour (section 5.).

4. Verification that the controllable behaviour designed using the control law satisfies the constraints of the control problem.

5. Rewriting of the given control law into a computer program. It can be done in a straightforward manner as the control laws determines directly the possible control actions.

In this way, a synthesis procedure for feedback controllers for a class of systems can be proved and effectively programmed [12, 13]. Of course, it is obvious that the design procedure can become computationally very expensive for some classes of systems. Therefore, the methods of problem decomposition and of behaviour approximation are investigated. This design methodology is used e.g. in [12, 13, 15].

Each control problem has usually several solutions which can be partially ordered with respect to set inclusion. A solution $\mathcal{Z}$ to the given control problem is called minimally restrictive if there does not exist any solution $\mathcal{Z}'$ such that $\mathcal{Z}$ is a strict subset of it.

Note that the minimally restrictive solutions contain also "the largest number of possible input trajectories". Therefore, they describe, in a certain sense, all ways in which the given system can be controlled while that the given constraints are satisfied. This concept is different from the concept of supremal controllable sublanguage where the order according to the logical output behaviour is considered.

7. STRONGLY CAUSAL SYSTEMS AND CONTROLLERS

The previous sections present results on causal systems and controllers. Analogous results can be proved also for strongly causal systems and controllers [10, 11]. They can be found very useful, as strong causality is one of the important features of real-world systems.

Let us consider a time set $(T, \leq)$ and sets of discrete event labels U and Y. Let $\Sigma = U \cup Y$ and $\mathcal{M} \subseteq \mathcal{F}(T, 2^{\Sigma})$. An output law $\mu_o : \mathcal{M} \times T \mapsto 2^{2^Y} \setminus \{\emptyset\}$ is strong if for all $f, g \in \mathcal{M}$ and $t \in T$ the following condition

$$m_U \circ f \,|\, \mathcal{J}_t^s = m_U \circ g \,|\, \mathcal{J}_t^s \ \wedge \ m_Y \circ f \,|\, \mathcal{J}_t^s = m_Y \circ g \,|\, \mathcal{J}_t^s \tag{9}$$

implies that $\mu_o(f, t) = \mu_o(g, t)$.

A TS $\mathcal{S} = (T, \leq, U, Y, \mathcal{G})$, where $\mathcal{G}$ is defined by (5), is strongly causal if condition (6) is satisfied for all $f \in \mathrm{pref}(\mathcal{M})$ [11].

Let us consider an arbitrary TS $\mathcal{S} = (T, \leq, U, Y, \mathcal{G})$ and a controllable behaviour $\mathcal{G}'$ of $\mathcal{S}$. Let $\mathcal{S}' = (T, \leq, U, Y, \mathcal{G}')$ be a TS. The set $\mathcal{G}'$ is a *strongly controllable behaviour* if for all $f \in \mathcal{G}'$, $A \in \mathcal{I}_T^b$, $u \in m_U \circ \mathcal{O}_{\mathcal{S}'}(f \mid A)$ and $y \in m_Y \circ \mathcal{O}_{\mathcal{S}'}^Y(m_Y \circ f \,|\, A)$ such that $\sup_{(T, \leq)}(A)$ exists the following condition

$$\exists g \in \mathcal{G}, \ m_U \circ g \,|\, \widetilde{A} = u \,|\, \widetilde{A} \ \wedge \ m_Y \circ g \,|\, \widetilde{A} = y \,|\, \widetilde{A}$$

implies that

$$\exists g \in \mathcal{G}', \ m_U \circ g \,|\, \widetilde{A} = u \,|\, \widetilde{A} \ \wedge \ m_Y \circ g \,|\, \widetilde{A} = y \,|\, \widetilde{A}.$$

In [10], it is proved in a constructive way, that there exists a strongly causal feedback controller $\mathcal{S}_2$ which realises a behaviour $\mathcal{G}'$ while interconnected with $\mathcal{S}_1$

if and only if $\mathcal{G}'$ is a strongly controllable behaviour of $\mathcal{S}_1$. Moreover it is shown (again in a constructive way), that if $\mathcal{S}_1$ is a TDES (ordinary or finitary ordinary TDES) than the controller can be also TDES (ordinary or finitary ordinary TDES, respectively).

Now we shall present how the strongly controllable behaviours can be constructed using strong control laws.

A control law $\mu_c : \mathcal{M} \times T \mapsto 2^{2^U} \setminus \{\emptyset\}$ is strong if for all $f, g \in \mathcal{M}$ and $t \in T$ condition (9) implies that $\mu_c(f, t) = \mu_c(g, t)$.

Let us consider a strongly causal ordinary TDES $\mathcal{S}' = (T, \leq, U, Y, \mathcal{M})$ with its behaviour constructed on the basis of a set $\mathcal{M}'$ and of an output behaviour μ_o' in the way presented above, $i.e.$ satisfying (6) if we put μ_o' for μ_o, $\mathcal{M}'$ for $\mathcal{M}$ and $\mathcal{M}$ for $\mathcal{G}$. The set $\mathcal{G}$ defined by (7) is a strongly controllable behaviour of $\mathcal{S}'$ if the condition (8) is fulfilled. A more general version of this result is proved in [11].

The results presented in this section give the possibility to use the same procedure for the design of strongly causal controllers as proposed in section 6. An application of these results is presented in [12, 15].

Note that we suppose only that the time set T is linearly ordered. Therefore, the results have a very general validity. They can be used for continuous time modelling as well as for discrete time modelling.

8. CONCLUSION

The key idea for the development of the new unifying DES framework [9, 10, 11, 12, 13, 14] is the consistent reformulation of fundamental concepts of system theory at the qualitative level. It gives also new and deeper insight into the concepts of system theory, e.g. causality [14].

The results are used for a real-time extension of a logical DES framework [12]. Certain preliminary results on control of systems with communication delays have been published in [13].

The main advantage of the proposed framework is that it is sufficiently general and at the same time it satisfies the requirements listed in section 1. Moreover, it is immediately applicable to the design of DES feedback controllers [12, 13, 15].

Current research is focused on general results concerning interconnections of systems and on control of systems with time forced events. Also different possibilities of development of a modular or hierarchical technique within the framework are investigated.

REFERENCES

[1] Collective. Challenges to control: A collective view. *IEEE Trans. Autom. Control,* 32(4):275–285, April 1987.

[2] P. Varaiya and H. Kurzhanski, editors. *Discrete Event Systems: Models and Application,* volume 103 of *Lecture Notes in Control and Information Sciences,* Berlin, Germany, August 1987. Springer Verlag. (IIASA Conference, Sopron, Hungary).

[3] Y. C. Ho, editor. Special issue on the dynamics of discrete event systems. *Proc. of the IEEE*, 77(1), January 1989.

[4] C. G. Cassandras, editor. *Workshop on Discrete Event Systems*, Amherst, MA, USA, June 1991. (Abstract Collection).

[5] M.D. Mesarovic and Y. Takahara. *General Systems Theory: Mathematical Foundations*. Academic Press, New York, 1975.

[6] M.D. Mesarovic and Y. Takahara. *Abstract Systems Theory*, volume 116 of *LNCIS*. Springer–Verlag, New York, 1989.

[7] J.C. Willems. Models for dynamics. In *Dynamics Reported*, New York, 1989. John Wiley & Sons.

[8] J.C. Willems. Paradigms and puzzles in the theory of dynamical systems. *IEEE Trans. Autom. Control*, 36(3):259–294, March 1991.

[9] P. Kozák. Discrete events and general systems theory. *Int. J. Systems Science*, 23(9):1403–1422, September 1992.

[10] P. Kozák. On feedback controllers. *Int. J. Systems Science*, 23(9):1423–1431, September 1992.

[11] P. Kozák. On controllable behaviours of time systems. Technical Report # 1739, Czechoslovak Academy of Sciences, Institute of Information Theory and Automation, Prague, Czechoslovakia, 1991. submitted.

[12] P. Kozák. Supervisory control of discrete event processes: A real-time extension. Technical Report # 1692, Czechoslovak Academy of Sciences, Institute of Information Theory and Automation, Prague, Czechoslovakia, 1991, revised 1992. submitted.

[13] P. Kozák. Control of elementary discrete event systems: Synthesis of controller with non-zero decision time. In D. Franke and F. Kraus, editors, *Proc. of the 1st IFAC Symposium on Design Methods of Control Systems*, pages 457–462, Zurich, Switzerland, September 1991. Pergamon Press, Oxford.

[14] P. Kozák. Causality and non-determinism. In R. Trappl, editor, *Proc. of the 11th European Meeting on Cybernetics and System Research*, pages 137–143, Vienna, Austria, April 1992. World Scientific Publishing Co. Pte. Ltd., Singapore.

[15] P. Kozák. The cat-and-mouse problem with least delays. *In this volume*, pages 199–206.

[16] G. Cohen, D. Dubois, J.P. Quadrat, and M. Viot. A linear-system-theoretic view of discrete-event processes and its use for performance evaluation in manufacturing. *IEEE Trans. Autom. Control*, 30(3):210–220, March 1985.

[17] P.J. Ramadge and W.M. Wonham. Supervisory control of a class of discrete event processes. *SIAM J. Control Optim.*, 25(1):206–230, January 1987.

Chapter II

Optimisation

SYNCHRONIZED CONTINUOUS FLOW SYSTEMS

Geert Jan Olsder*

Abstract. Continuous analogues of the descriptions of discrete event dynamic systems and/or timed Petri nets will be given. Thus it will be shown that the synchronization aspects of such systems are more basic than their discrete character. Various theorems in the context of discrete event systems have their logical counterparts within the continuous setting.

1. INTRODUCTION

During the last decade or so there have been many scientific activities in the area of discrete event dynamic systems (DEDS) and they will most likely continue for quite some time. One of the reasons for these activities is the increasing importance of parallel and distributed processing in modern technology. One of the striking features of DEDS is that the evolution of the system is not explicitly determined by a global clock (which is the case if one deals with difference or differential equations), but rather by events inside the system itself. New modeling and mathematical techniques are being developed for DEDS. The terminology 'discrete event dynamic systems' already betrays the discrete character of the problems studied within the context of this theory. Examples of such events are the arrival of some data, the closing of a gate, etc.

A particular class of DEDS is based on the so-called max-plus algebra, which means that the operations in the underlying algebra for the systems studied are maximization and addition. Within this class of systems one studies the explicit time behavior. One sometimes speaks of timed DEDS. Typical items in this context are performance analysis and throughput. These DEDS form an analytical basis for studying timed event graphs. The set of (timed) event graphs is a proper subset of (timed) Petri nets.

Recently there have been attempts to consider 'continuous' versions of timed Petri nets or timed event graphs, see [1], [2]. These continuous versions have been obtained by limiting arguments when the discretization (both with respect to time and tokens) was made smaller and smaller. These papers did not consider positive transportation times between the nodes (nodes are called 'transitions' in the Petri net terminology). Reference [2] states explicitly that in Petri nets delays in transitions can be transformed to delays in places and vice versa. This statement is questionable for general Petri nets. It is used, however, as an argument to only deal with timed transitions and not with timed transportation. The current paper does consider such positive transportation times. It is surprising to realize that many results in the discrete setting do have their counterparts in

*Department of Technical Mathematics and Informatics, Delft University of Technology, P.O. Box 5031, 2600GA Delft, the Netherlands; e-mail: witagjo@dutinfh.tudelft.nl

the continuous setting. Therefore there are DEDS and CEDS (Continuous Event Dynamic Systems) – though the latter may not be a good terminology – as there are difference and differential equations. If a new acronym must be coined, SCFS for Synchronized Continuous Flow Systems would be more appropriate. But then DEDS should be replaced by SDFS, which stands for Synchronized Discrete Flow Systems. Anyhow, the synchronization aspect, a transition works as fast as the minimum incoming flow allows (it synchronizes the incoming flows), seems to be the basic feature of these systems and not whether the flows are discrete or continuous.

A possible advantage of working with SCFS is the following. If one has to deal with a large timed Petri net, its mathematical model will have a large size (in terms of the dimension of the state for instance). Approximating the discrete flows in this Petri net by continuous ones might lead to a model of smaller size.

2.   TIMED PETRI NETS

It is assumed that the reader is familiar with the basic properties of Petri nets, see [3]. In order to set the notation and the stage, we do start with some formal definitions though.

Definition 1 (Petri net) *A Petri net is a pair $(\mathcal{G}, b)$, where $\mathcal{G} = (\mathcal{V}, \mathcal{E})$ is a bipartite graph with a finite number of nodes (the set $\mathcal{V}$) which are partitioned into the disjoint sets $\mathcal{P}$ and $\mathcal{Q}$; $\mathcal{E}$ consists of pairs of the form (p_i, q_j) and (q_j, p_i) with $p_i \in \mathcal{P}$ and $q_j \in \mathcal{Q}$. The initial marking b is an m-vector, with m being the number of elements in $\mathcal{P}$, of nonnegative integers. The elements of $\mathcal{P}$ are called places, those of $\mathcal{Q}$ are called transitions. The number of elements in these sets are m and n respectively. The elements of the vector b denote the number of tokens in the respective places.*

Tokens represent the 'flow' in the network. Tokens move through the network if transitions 'fire'. At such a firing a transition takes one token of all of its upstream places, provided all these places contain at least one token which is enabled. If not each of these places contains an enabled token, then the transition can not fire. After the firing has ended one token is put in each of the downstream places. Such a token becomes enabled again if it has spent a certain time in this place.

Definition 2 (Event graph) *A Petri net is called an event graph if each place has exactly one upstream and one downstream transition.*

Definition 3 (Firing time) *The firing time of a transition is the time that elapses between the starting and the completion of the firing of the transition.*

Definition 4 (Holding time) *The holding time of a place is the time a token must spend in the place before it can contribute to the enabling of the downstream transitions.*

Theorem 1 *An event graph with both firing times and holding times is equivalent to an event graph with only holding times (i.e. the firing times are zero). This equivalence means that the time instants at which the transitions fire are the same in both event graphs.*

Proof: See [4]. ∎

From now on we will only consider event graphs with firing times which are zero. Each place connects precisely one transition with precisely one (possibly different) transition. One says in such a situation that the upstream transition, say q_j, is a predecessor of the downstream transition, say q_i. Equivalently one can say that q_i is a successor of q_j. One writes in such a case $j \in \pi^-(i)$ and $i \in \pi^+(j)$.

We make the explicit assumption that if a place connects two transitions such as just has been described, there is no other place with does exactly the same. In general event graphs there can be more 'parallel' places in between two transitions of which one is the successor of the other. The reason for this restriction is purely a notational issue. The theory to be given can handle the more general situation routinely. We also make the assumption that the underlying network is strongly connected.

If a place exists between the transitions q_j and q_i and q_j is upstream with regard to this place and q_i downstream, then the holding time of this place is indicated by a_{ij}. The holding times are nonnegative real numbers. The number of tokens in this place is indicated by b_{ij}.

If $x_i(t)$ denotes the number of firings of transition q_i, $i = 1, \ldots, n$, which have taken place up to, and including, time t, one gets the equations

$$x_i(t) = \bigoplus'_{j \in \pi^-(i)} b_{ij} \otimes x_j(t - a_{ij}), \ i = 1, \ldots, n. \tag{1}$$

These are the so-called counter equations. The symbol $\bigoplus'$ refers to minimization and the symbol $\otimes$ refers to addition. Hence in more conventional writing (1) would become

$$x_i(t) = \min_{j \in \pi^-(i)} b_{ij} + x_j(t - a_{ij}), \ i = 1, \ldots, n. \tag{2}$$

Dual to these equations one has the so-called dater equations;

$$\tau_i(\chi) = \bigoplus_{j \in \pi^-(i)} a_{ij} \otimes \tau_j(\chi - b_{ij}), \ i = 1, \ldots, n. \tag{3}$$

The symbol $\bigoplus$ refers to maximization. The quantity $\tau_i(\chi)$ denotes the earliest time instant at which transition q_i has fired χ times. Note that χ and x_i are integer-valued. The functions $x_i(t)$ and $\tau_i(\chi)$ are each others inverse in a way. Both (1) and (3) are (different) descriptions of the same underlying system. With the conventional way of writing (3) would become

$$\tau_i(\chi) = \max_{j \in \pi^-(i)} a_{ij} + \tau_j(\chi - b_{ij}), \ i = 1, \ldots, n. \tag{4}$$

One can make the resemblance between x_i and τ_i more striking if one assumes that a quantity $g_c > 0$ exists such that all a_{ij} values are integer multiples of this quantity. If so, one can scale the time in such a way that g_c becomes the new time unit. If one does so, all the quantities x_i, τ_i, t and χ become integer-valued. If $a_{ij} > 0$, i.e. $a_{ij} \in \{1, 2, \ldots\}$, for all combinations i and j such that $j \in \pi^-(i)$, then (1) has a recursive character. With the appropriate initial conditions these equations can be solved recursively. If one or more of these quantities a_{ij} are zero, (1) is an implicit equation, for which a solution may or may not exist. Similar remarks hold with respect to (3) if $b_{ij} = 0$ for some admissible combination of i and j. The issue of initial conditions for either (1) or (3) will not be addressed here explicitly; see [4] for a detailed discussion.

If in the original event graph there would have been a positive firing time, then the equations above do not exclude the possibility that a transition 'works' simultaneously on two or more tokens. If one wants to exclude this, a loop, including a place, around the transition concerned should be added. The holding time of this new place is defined to be equal to the original firing time of this transition. This loop now takes care of the fact that in the equivalent event graph with only zero firing times, the transition cannot work on two or more tokens simultaneously anymore. A possible interpretation of such a loop is that the transition has a capacity constraint. We will come back to this issue in section 4.

3. CONTINUOUS FLOWS, NO CAPACITY CONSTRAINTS

The central equations of this section are (1) and (3). It is assumed now that in addition to t and τ_i, also χ and x_i are real-valued, and so are the quantities b_{ij}. The interpretation of these equations is still a (strongly connected) network with n transitions (also called nodes now). These nodes can now fire continuously. The intensity of this firing is indicated by $v_i(t)$. Quantity $x_i(t)$ denotes again the total amount produced by node i up to (and including) time t. As initial condition it is assumed that $x_i(0) = 0$. The production of a continuously firing transition is sent with unit speed along the outgoing arcs to the downstream transitions. Thus along an arc there is a continuous flow. The intensity of this flow is $\varphi_i(t, l)$, where l is the parameter indicating the exact location along an arc starting from q_i; $l = 0$ coincides with the beginning of such an arc, $l = a_{ji}$ coincides with the end of this arc, where it is assumed that the downstream transition is q_j. As long as the parameters lie in appropriate intervals, we have $\varphi_i(t, l) = \varphi_i(t + s, l + s)$. Moreover, $\varphi_i(t, l) = \varphi_i(t - l, 0) = v_i(t - l)$.

At time t the total amount of material along the arc from q_i to q_j equals

$$\int_{l=0}^{l=a_{ji}} \varphi_i(t, s)\, ds. \tag{5}$$

The quantities b_{ij} satisfy $b_{ij} = \int_0^{a_{ij}} \varphi(0, s)\, ds$. The integrand and the integral in (5) must be considered with some care. It is quite well possible that the integrand contains a δ- function. This will particularly happen at the end of an arc, when material must wait there to be processed by the downstream transition because the other incoming arcs to the same transition have brought in less material sofar. If q_k is a downstream transition to both q_i and q_j and if $x_i(t) < x_j(t)$, then φ_j will start to build a δ-function at $l = a_{kj}$, from t onwards. Of course this δ-function can disappear again later on if $x_i(s) > x_j(s)$ for an s-value with $s > t$. The total amount of material along an arc, as expressed by (5), will in general be time dependent. However, we have the following theorem.

Theorem 2 *Along a circuit the total amount of material is constant; if the circuit ζ is characterized by the transitions $\{q_{i_1}, q_{i_2}, \ldots, q_{i_{k+1}} = q_{i_1}\}$, then*

$$\sum_{l=1}^{l=k} \int_0^{a_{\pi^+(i_l),i_l}} \varphi_{i_l}(t, s)\, ds$$

is constant (it does not depend on time).

Proof: A firing transition takes from every incoming arc exactly as much material as it puts in each of the outgoing arcs. ∎

Please note that the total amount of material in the network is not necessarily constant.

Definition 5 (Cycle mean) *Given a circuit $\zeta = \{q_{i_1}, q_{i_2}, \ldots, q_{i_{k+1}} = q_{i_1}\}$, its weight $]\zeta]_w$ and its length $]\zeta]_l$ are defined as*

$$]\zeta]_l = \sum_{l=1,\ldots,k} b_{\pi^+(i_l),i_l},$$

$$]\zeta]_w = \sum_{l=1,\ldots,k} a_{\pi^+(i_l),i_l}.$$

If $]\zeta]_l > 0$, the cycle mean is defined as $]\zeta]_w/]\zeta]_l$.

It will be assumed that $]\zeta]_l > 0$ for all circuits. The reason for this assumption is that if the total amount of material in a circuit would be zero, it will remain zero forever due to Theorem 2 and the transitions in this circuit will remain idle forever. For the Equations (1) to be solvable unambiguously, it must be assumed that all a_{ij}-parameters are strictly positive. If this is true, (1) determines the future evolution of the x_i's uniquely, provided appropriate initial conditions are given. These initial conditions (at time $t = 0$) consist of $x_i(0) = 0$ and of some history of the x_i-values for negative t values (imagine that the system has been working already before time $t = 0$). The running of the system before $t = 0$ has lead to a distribution of the material (the 'smashed' tokens) along the arcs at $t = 0$. Hence an equivalent set of initial conditions is formed by the x_i values together with the distribution of the material along the arcs, all at $t = 0$.

Assumptions. The following assumptions will hold for all theorems to be formulated in this and the next section:

- the network is strongly connected;

- maximally one arc exists between two nodes;

- a_{ij} and b_{ij} are nonnegative;

- $]\zeta]_l > 0$ and $]\zeta]_w > 0$ for all circuits ζ.

Definition 6 (Critical circuit) *The circuits which have the maximum cycle mean are called critical. The corresponding cycle mean is indicated by λ.*

Theorem 3 *For suitably chosen initial conditions, Equations* (1) *have a solution*

$$x_i(t) = \frac{1}{\lambda}t + d_i,$$

where d_i are constants.

Proof: Suppose that solutions of the form $x_i(t) = e_i t + d_i$ exist. They are then substituted into (1) leading to the identities

$$e_i t + d_i = \min_{j \in \pi^-(i)} b_{ij} + e_j((t - a_{ij}) + d_j), \ i = 1, \ldots, n. \tag{6}$$

For large values of t this leads to

$$e_i = \min_{j \in \pi^-(i)} e_j, \ i = 1, \ldots, n.$$

Due to the assumption that the network is strongly connected, all e_i values must be equal; this value will be denoted e. If we now substitute $t = 0$ into (6), the result is

$$d_i = \min_{j \in \pi^-(i)} b_{ij} - e a_{ij} + d_j, \ i = 1, \ldots, n,$$

which can be written in min-plus notation as

$$d = Rd,$$

where $d = (d_1, \ldots, d_n)'$, the symbol $'$ denoting transposed, and where the (i,j)-th element of the matrix R equals $b_{ij} - e a_{ij}$, provided that $j \in \pi^-(i)$, otherwise this element equals $+\infty$. The vector d is an eigenvector of R, corresponding to the eigenvalue zero (in the min-plus algebra sense). The remaining question is whether e can be chosen such that R has an eigenvalue zero. Due to the strong connectedness assumption again, R is irreducible and hence it has only one eigenvalue which equals the minimum cycle mean (recall that we are working in the min-plus algebra now; the definition of minimum cycle mean will be obvious)

$$\min_\zeta \frac{\sum_{(i,j) \in \zeta}(b_{ij} - e a_{ij})}{]\zeta]_1} = \min_\zeta \frac{\sum_{(i,j) \in \zeta} b_{ij} - e \sum_{(i,j) \in \zeta} a_{ij}}{]\zeta]_1}.$$

If we choose

$$e = \min_\zeta (\sum_{(i,j) \in \zeta} b_{ij}) / (\sum_{(i,j) \in \zeta} a_{ij}) = \frac{1}{\max_\zeta (\sum_{(i,j) \in \zeta} a_{ij}) / (\sum_{(i,j) \in \zeta} b_{ij})} = \lambda^{-1},$$

then this minimum cycle mean becomes zero as required. This concludes the proof. ∎

Please note that the density functions $\varphi_i(t,s)$ are constant; $\varphi_i(t,s) = e$, with a possible exception of the endpoints of the corresponding arcs, where a δ-function may be present (of which the intensity is constant with time again).

The solution given in this proof is, in the sense to be given, the 'best' one. Let us concentrate on the critical circuit (or, if there are more, on a critical circuit).

During a time interval of $\sum a_{ij}$ time units, where the summation is over all arcs of this circuit, any transition within this critical circuit can never produce more than an amount of $\sum b_{ij}$, the summation being again over all arcs of the circuit. In the linear solution given above any transition of the critical circuit produces exactly an amount of material equal to $\sum b_{ij}$ in each of the outgoing arcs during $\sum a_{ij}$ time units.

Example 1 Other solutions to (1) can exist. By means of an example it will be shown that a solution exists which fluctuates in a periodic way around the linear solution obtained in the theorem above. We are given a network with $n = 2$. It is assumed that all four arcs (from transition q_i to q_j, $i, j = 1, 2$) exist and that

$$\lambda \stackrel{\text{def}}{=} \frac{a_{12} + a_{21}}{b_{12} + b_{21}} > \frac{a_{ii}}{b_{ii}}, \, i = 1, 2. \tag{7}$$

We try a solution of the form

$$x_i(t) = \lambda^{-1}t + \alpha_i \sin(\beta(t - r_i)) + s_i, \, i = 1, 2$$

with $r_2 = s_2 = 0$ and

$$]\alpha_i]\beta < \lambda^{-1}, \, i = 1, 2. \tag{8}$$

The latter two conditions ensure the solutions, if they exist, to be nondecreasing. This solution is substituted into (1), leading to the identities

$$\begin{aligned}
s_1 + \alpha_1 \sin(\beta(t - r_1)) &= \min\{-\lambda^{-1}a_{11} + \bar{b}_{11} + \alpha_1 \sin(\beta(t - a_{11} - r_1)), \\
&\quad -\lambda^{-1}a_{12} + b_{12} + \alpha_2 \sin\beta(t - a_{12})\}, \\
\alpha_2 \sin(\beta t) &= \min\{-\lambda^{-1}a_{21} + \bar{b}_{21} + \alpha_1 \sin(\beta(t - a_{21} - r_1)), \\
&\quad -\lambda^{-1}a_{22} + b_{22} + \alpha_2 \sin\beta(t - a_{22})\},
\end{aligned} \tag{9}$$

where $\bar{b}_{i1} = s_1 + b_{i1}$. Each minimization operation has two arguments; if we assume for the moment that these arguments satisfy

$$\begin{aligned}
-\lambda^{-1}a_{11} + s_1 + b_{11} + \alpha_1 \sin(\beta(t - a_{11} - r_1)) &\geq \\
-\lambda^{-1}a_{12} + b_{12} + \alpha_2 \sin\beta(t - a_{12}), \\
-\lambda^{-1}a_{21} + s_1 + b_{21} + \alpha_1 \sin(\beta(t - a_{21} - r_1)) &\leq \\
-\lambda^{-1}a_{22} + b_{22} + \alpha_2 \sin\beta(t - a_{22}),
\end{aligned} \tag{10}$$

then the identities become

$$\begin{aligned}
s_1 + \alpha_1 \sin(\beta(t - r_1)) &= -\lambda^{-1}a_{12} + b_{12} + \alpha_2 \sin\beta(t - a_{12}), \\
\alpha_2 \sin(\beta t) &= -\lambda^{-1}a_{21} + s_1 + b_{21} + \alpha_1 \sin(\beta(t - a_{21} - r_1)).
\end{aligned} \tag{11}$$

These equations are indeed identities if

$$\begin{aligned}
r_1 &= a_{12}, \tag{12} \\
\beta(a_{12} + a_{21}) &= 2k\pi, \, k = 1, 2, \ldots, \tag{13} \\
\alpha_1 &= \alpha_2, \tag{14} \\
s_1 &= -\lambda^{-1}a_{12} + b_{12}, \tag{15} \\
-\lambda^{-1}a_{21} + s_1 + b_{21} &= 0. \tag{16}
\end{aligned}$$

The quantities s_1 and λ can be uniquely solved from (15) and (16). The value of λ fortunately coincides with its value given in (7). The value β is determined by (13). Thus it is shown that a 'periodic' solution exists, provided that (8) and (10) are true. A simple analysis shows that this is the case for $]\alpha_1](=]\alpha_2])$ sufficiently small. For $k = 1$ in (13) we get β equals the length of the critical circuit divided by 2π. For larger values of k we get higher harmonics. If the results of various k values are combined, the solution becomes a Fourier series (with period $a_{12} + a_{21}$) added to the linear part.

Depending on the parameters a_{ij} and b_{ij}, one can construct a Fourier series in such a way that the ultimate solution $x_i(t)$ becomes piecewise constant (and nondecreasing), thus leading to a *real* discrete flow again. Points at which $x_i(t)$ jumps refer to discrete events.

Conjecture 1 *If there is a unique critical circuit, then each solution of (1), starting from arbitrary initial conditions, converges in finite time either to the linear solution or to a 'periodic' solution as described in the example just given.*

The proof of this conjecture could resemble the train of thought of the proof of the discrete analogue of this theorem, which can be found in [5]. A sketch of the proof is as follows. Given the value $x_i(t)$ for some t sufficiently large, a critical path along the nodes, backward in time, is constructed according to (1), leading back all the way to the initial conditions. This critical path will contain a number of encirclements of the critical circuit, denoted ζ_{crit}. If now the same is done with respect to $x_i(t+]\zeta_{crit}]_w)$, one gets the same critical path, except for the fact that ζ_{crit} will be encircled once more than for the critical path corresponding to $x_i(t)$. Therefore $x_i(t+]\zeta_{crit}]_w) = x_i(t)+]\zeta_{crit}]_l$.

Another way to possibly prove the conjecture above is to discretize the time units and material units (the tokens) in small units and apply the proof of the discrete version of this theorem. Subsequently the discretization is made smaller and smaller and then a continuity argument is invoked.

If the critical circuit is nonunique, remarks related to the periodicity similar to those in [5] can be given for the continuous case.

4. CONTINUOUS FLOWS WITH CAPACITY CONSTRAINTS

The basic formulas in this section are

$$v_i(t) = \begin{cases} c_i, & \text{if } m_{ij}(t) > 0 \text{ for all } j \in \pi^-(i), \\ c_i \oplus' \bigoplus'_{j\in\pi^-(i)} v_j(t - a_{ij}), & \text{otherwise}, \end{cases}$$
$$\dot{m}_{ij}(t) = v_j(t - a_{ij}) - v_i(t), \tag{17}$$

for $i = 1, \ldots, n$. The quantity v_i denotes the production flow, i.e. the production per time unit, of transition q_i. The relation with the total amount produced is

$$x_i(t) = \int_{s=0}^{s=t} v_i(s)\, ds.$$

The quantity $m_{ij}(t)$ represents the intensity of the δ-function at the end of the arc from q_j to q_i, as described in the previous section. It is easily checked that

$m_{ij}(t) \geq 0$ for $t \geq 0$, provided that the initial condition $m_{ij}(0)$ is nonnegative. It may look somewhat surprising that the quantities b_{ij} have disappeared from (17). They are, however, implicitly present in the initial conditions; in order to calculate $v_i(t)$ for $t > 0$, one needs 'old' values of $v_i(t)$ with $t < 0$, which are related to the quantities b_{ij}.

Please note that (17) can not be obtained by taking finer and finer discretizations (with respect to both time and tokens) as done in [1]. If one would start with a loop with a place around a transition in the discrete setting in order to express the finite capacity constraint of this transition, then by taking finer and finer discretizations, this loop becomes an arc with positive transportation time in the limit. The proper way to view the capacity constraint as a limit of a loop in a Petri net is still to make the discretization of both time and material smaller and smaller, but the number of time units and the amount of material in the loop remains constant, i.e. independent of the discretization used. Such is for instance the holding time of the loop always one. In the limit as the discretization approaches zero units, the ratio of matrial and the holding time in the loop becomes the capacity constraint. In 'conventional' arcs, i.e. those which are not introduced for the purpose of capacity constraints, the number of time units and the amount of material do increase as the discretization becomes finer and finer.

The equivalent expressions of (17) in the dater sense are

$$
v_i(\chi) = \begin{cases} \frac{1}{c_i}, & \text{if } \mu_{ij}(\chi) > 0 \text{ for all } j \in \pi^-(i), \\ \frac{1}{c_i} \oplus \bigoplus_{j \in \pi^-(i)} v_j(\chi - b_{ij}), & \text{otherwise,} \end{cases}
$$
$$
\mu_{ij}(\chi) = v_j(\chi - b_{ij}) - v_i(\chi),
$$

(18)

for $i = 1, \ldots, n$.

Theorem 4 *Along a circuit the total amount of material is constant.*

The proof is identical to that of Theorem 2.

In the following theorem the quantity λ appears again. As in Section 3, it is equal to the maximum cycle mean. The definition of the cycle mean, however, must now slightly be adapted.

Definition 7 (Cycle mean) *Given a circuit $\zeta = \{q_{i_1}, q_{i_2}, \ldots, q_{i_{k+1}} = q_{i_1}\}$, the weight $]\zeta]_w$ and the length $]\zeta]_l$ are defined as*

$$
]\zeta]_l = \sum_{l=1,\ldots,k} b_{\pi^+(i_l),i_l} + m_{\pi^+(i_l),i_l}(0),
$$

$$
]\zeta]_w = \sum_{l=1,\ldots,k} a_{\pi^+(i_l),i_l}.
$$

If $]\zeta]_l > 0$, the cycle mean of ζ is defined as $]\zeta]_w /]\zeta]_l$.

The term $m_{\pi^+(i_l),i_l}(0)$ in this definition refers to a concentration of material at the end of the arc from q_{i_l} to $q_{\pi^+(i_l)}$. It is understood here that the function $\varphi(0, s)$ is a 'real' one and does not contain δ-functions anymore. Because of the capacity constraints $\varphi(t, s)$ will also be a 'real' function for $t > 0$.

Theorem 5 *For appropriately chosen initial conditions, Equations (17) have a solution* $x_i(t) = \overline{\lambda}t + \overline{d}_i$, *where* $\overline{\lambda} = \min(c_1, \ldots, c_n, \lambda^{-1})$.

Proof: If $\lambda^{-1} \leq \min_i c_i$, the proof is identical to the one of Theorem 5. If $\lambda^{-1} > \min_i c_i$, the assertion of the theorem follows from direct substitution of the proposed solution into (17). ■

Conjecture 2 *If there is a unique critical circuit, then each of* (17), *starting from arbitrary initial conditions, converges in finite time either to the linear solution or to a 'periodic' one.*

The proof is probably almost identical to the proof of Theorem 1. The only difference is with respect to the discretization of the loops added to represent the capacity constraints.

Example 2 Consider a network with three nodes. The transportation times are $a_{21} = 2$, $a_{12} = 2$, $a_{13} = 2$, $a_{31} = 5$, $a_{32} = 4$, $a_{33} = 3$. The time durations a_{ij} which have not been mentioned refer to nonexisting arcs. The capacity constraints are $c_1 = 2$, $c_2 = 3$, $c_3 = 4$. The initial values of the flows, $\varphi_i(0, s)$, $0 \leq s \leq a_{ji}$, for $i = 1, 2, 3$ and the appropriate j, are piecewise constant;

$$
\begin{aligned}
&\varphi_1(s) = 1, \quad \text{for } 0 \leq s \leq 1; \quad \varphi_1(s) = 2, \quad \text{for } 1 < s \leq 2; \\
&\varphi_1(s) = 3, \quad \text{for } 2 < s \leq 3; \quad \varphi_1(s) = 1, \quad \text{for } 3 < s \leq 4; \\
&\varphi_2(s) = 1, \quad \text{for } 0 \leq s \leq 1; \quad \varphi_2(s) = 2, \quad \text{for } 1 < s \leq 2; \\
&\varphi_2(s) = 2, \quad \text{for } 2 < s \leq 3; \quad \varphi_3(s) = 1, \quad \text{for } 0 \leq s \leq 1; \\
&\varphi_3(s) = 1, \quad \text{for } 1 < s \leq 2.
\end{aligned}
\tag{19}
$$

This list of φ-values only gives the flow intensities along the arcs from q_j to q_i for $0 \leq s \leq a_{ij} - 1$. The reason that this list is not meant for the last parts of the arcs is the following. The initial masses at the end of the arcs are, contrary to (17), not concentrated in one time point (as a δ-function), but as masses at the last discretization interval of the arcs in this example. They are:

$$
\begin{aligned}
&m_{21}(0) = 4, \quad m_{12}(0) = 1, \quad m_{31} = 1, \\
&m_{13}(0) = 2, \quad m_{32}(0) = 0, \quad m_{33} = 3.
\end{aligned}
\tag{20}
$$

To be clear, the different levels of flow along the arc from q_1 to q_2, for instance, are 1, 4. Thus all the necessary initial conditions for (17) are given and its future behavior can be calculated. The advantage of the initial piecewise constant flow distributions along the arcs is that these flow distributions, though continuously moving along the arcs, remain piecewise constant. The points of discontinuity move with a speed of one unit downstream along the arcs. The lengths of the intervals between the points of discontinuity (except for the intervals at the beginning and the end of an arc obviously) equals g_c, which was introduced in section 2. In the current example $g_c = 1$. Rather then as a continuous event system, the current example can also be viewed as a discrete event system; this discrete event system becomes a continuous one by introduction of the so-called hold-operator which is used in D/A convertors.

The reason for the introduction of the special initial conditions is that calculations can easily be performed. If one confines oneself to integer values of the time, then by means of a simple program the evolution can be determined. There are four elementary circuits in the network and their cycle means are 4/7, 7/11, 8/13 and 3/5. Hence $\lambda = 7/11$. There is one critical circuit, viz. $1 \rightarrow 3 \rightarrow 1$. Since $(7/11)^{-1}$ is smaller than the minimum capacity constraint, we expect solutions to exist with an average throughput of 11/7. This is indeed true. from $t = 6$ onwards, the behavior is periodic with a period of 7; $v_i(t) = v_i(t+7)$, $m_{ij}(t) = m_{ij}(t+7)$ for $t \geq 6$. During one period, the transitions fire with intensities 2, 2, 1, 1, 2, 2 and 1 respectively, such that indeed the average throughput is 11/7.

The example given above is in a sense insipid since it is close to an event graph which behaves in the same way. The following example, however, will have the masses m_{ij} concentrated in one time point. It will be shown that these masses rather than being δ-functions at the end of an arc, can equally well be added to the material of the last part of the arc.

Example 3 This example is a variation of Example 2 in the sense that now the masses at the end of the arcs are concentrated in one time point. This means that now the transportation times are $a_{21} = 1$, $a_{12} = 1$, $a_{13} = 1$, $a_{31} = 4$, $a_{32} = 3$, $a_{33} = 2$. All other quantities which determine the problem remain the same. The cycle means now are 4/8, 5/11, 5/13, 2/5, such that $\lambda = 5/11$. Because of the constraint $v_1 = 2$ we expect that $\overline{\lambda} = 2$, which turns out to be true. In order to perform the calculations for the future behavior, the concentrated masses can be added to the material already present in the last part of the arc. Thus one obtains an example which has the same features as the previous one. Because of the piecewise constant material distributions, the solution is equally 'insipid'. The solution is now linear (no 'periodic' part) due to the fact that a capacity constraint determines the speed of the flows and such a capacity constraint has 'length zero'. The linear solution is reached after 7 time units in this example.

5. CONCLUSION

It has been shown that there are many similarities between discrete event systems and the newly introduced continuous counterparts of such systems.

REFERENCES

[1] R. David and H. Alla. Continuous Petri nets. In *Proceedings of the 8th European workshop on Application and Theory of Petri Nets, Saragossa, Spain*, pages 275–294, 1987.

[2] R. David and H. Alla. Autonomous and timed continuous Petri nets. Technical report, Laboratoire d'Automatique de Grenoble, ENSIEG, Saint-Martin-d'Hères, 1989.

[3] T. Murata. Petri nets: Properties, analysis and applications. *Proceedings of the IEEE*, 77:541–580, 1989.

[4] F. Baccelli, G. Cohen, G.J. Olsder, and J.P. Quadrat. *Synchronization and Linearity*. Wiley, 1992.

[5] G. Cohen, D. Dubois, J.P. Quadrat, and M. Viot. A linear-system theoretic view of discrete-event processes and its use for performance evaluation in manufacturing. *IEEE Transactions on Automatic Control*, AC-35:210–220, 1985.

ON A GENERALIZED ASYMPTOTICITY
PROBLEM IN MAX ALGEBRA

J. G. Braker*   J. A. C. Resing[†]

Abstract. The asymptotic behaviour of an algorithm in max algebra is discussed.
It can be seen as an extension of the concept of periodicity which has been treated
in max-algebra literature. The main result is that the asymptotic behaviour of the
algorithm is characterized by one or more critical circuits in a generalized sense,
and that two cases can be distinguished. In the first case the generalized critical
circuit has length two, and generalized order-2 periodicity is found. In the second
case the generalized critical circuits are of length one, and generalized order-1
periodicity is stated. Only the case of 2×2-matrices is treated. The treatment of
this specific case leads to the formulation of a more general set-up in which square
matrices of any size can be included. The notion of generalized critical circuit is
introduced.

1. MOTIVATION AND PRELIMINARIES

Example 1 *Consider a production network with the following description. The
network contains n nodes. We are interested in the time instant at which node
i ($1 \leq i \leq n$) becomes active for the kth time. This time instant will be denoted
by $x_i(k)$. Node i can start its $(k + 1)^{\text{st}}$ activity immediately after all nodes have
completed their k^{th} activity and supplied node i. Let $m_{ij}(k)$ denote the total time
(production plus transportation) between the start of the kth activity at node j and
the arrival of the corresponding supply at node i. Then the system is described by
the equations*

$$x_i(k + 1) = \max_{1 \leq j \leq n} (x_j(k) + m_{ij}(k)). \tag{1}$$

For the analysis of systems as described in (1) it is useful to introduce the max-
plus algebra, often just called max algebra. This max algebra is defined by
the operations addition and multiplication, applied to the set of real numbers
extended with $-\infty$. A suggestive notation is used for these operations. This
notation expresses the large extent of equivalence between max algebra and linear
algebra.

Definition 1 *The max algebra $(S, \otimes, \oplus)$ is composed of*

- $S = \mathbb{R} \cup \{-\infty\}$, *where $\mathbb{R}$ is the set of real numbers,*
- $\otimes$ *is addition in S, where $a \otimes -\infty = -\infty$ for all $a \in S$,*
- $\oplus$ *is maximization in the usual ordering of S.*

*Department of Technical Mathematics and Informatics, Delft University of Technology,
P.O. Box 5031, 2600 GA Delft, The Netherlands; e-mail: witajgb@dutinfh.tudelft.nl
†Department of Mathematics and Computing Science, Eindhoven University of Technology,
P.O. Box 513, 5600 MB Eindhoven, The Netherlands; e-mail: resing@bs.win.tue.nl

The max algebra is not a ring, because the operation $\oplus$ does not have an inverse. For instance, the equation $a \oplus x = b$ does not have a solution x if $a > b$. The structure $(S, \otimes, \oplus)$ is a semiring, also called dioid. See [1, 2, 3] for the theory of dioids.

As an example, we give below some calculations in max algebra:

$$1 \otimes 4 = 7, \quad 3 \oplus 4 = 4, \quad (3 \otimes 4) \oplus (2 \otimes 6) = 8, \quad (3 \oplus 4) \otimes (2 \oplus 6) = 10.$$

Clearly, the neutral element of S with respect to $\otimes$ ('one element') is 0, since $a \otimes 0 = a$ for all $a \in S$. The neutral element with respect to $\oplus$ ('zero element') is $-\infty$, since $a \oplus -\infty = a$ for all $a \in S$.

The above examples do not yet justify the notations $\oplus$ and $\otimes$. This justification comes from the introduction of matrix operations $\oplus$ and $\otimes$. Suppose the two equations

$$\begin{cases} y_1 &= \max(m_{11} + x_1, m_{12} + x_2), \\ y_2 &= \max(m_{21} + x_1, m_{22} + x_2) \end{cases} \tag{2}$$

hold. These equalities can be written in max algebra as

$$\begin{cases} y_1 &= (m_{11} \otimes x_1) \oplus (m_{12} \otimes x_2), \\ y_2 &= (m_{21} \otimes x_1) \oplus (m_{22} \otimes x_2). \end{cases} \tag{3}$$

Note the analogy to linear algebra as we write

$$\begin{bmatrix} y_1 \\ y_2 \end{bmatrix} = \begin{bmatrix} m_{11} & m_{12} \\ m_{21} & m_{22} \end{bmatrix} \otimes \begin{bmatrix} x_1 \\ x_2 \end{bmatrix}. \tag{4}$$

The above notation is the inspiration of the definition of matrix multiplication in max algebra.

Definition 2 *Let A and B be matrices of dimensions $m \times r$ and $r \times n$, respectively. Then for all $i = 1, \ldots, m$ and $j = 1, \ldots, n$ the* product $A \otimes B$ *is defined by*

$$(A \otimes B)_{ij} = \max_{k=1,\ldots,r} (A_{ik} + B_{kj}) = \sum_{k=1}^{r} {}_{\oplus} (A_{ik} \otimes B_{kj}). \tag{5}$$

In the sequel we shall write $M_{\otimes}^k$ for $M \otimes \cdots \otimes M$ (k times).

Definition 3 *A vector $v \neq [-\infty, \ldots, -\infty]^T$ and a scalar λ satisfying $M \otimes v = \lambda \otimes v$ are called* eigenvector *of M and* eigenvalue *of M, respectively.*

Definition 4 *The graph $\mathcal{G}_M$ corresponding to an $n \times n$-matrix M is a pair $(\mathcal{V}, \mathcal{E})$ where $\mathcal{V}$ is the set of nodes (vertices) of $\mathcal{G}_M$ and $\mathcal{E}$ is the set of arcs (edges) of $\mathcal{G}_M$. The nodes are numbered $1, \ldots, n$. Arc (j, i) from node j to node i is present in $\mathcal{G}_M$ if and only if $M_{ij} > -\infty$; in that case M_{ij} is the* weight *of arc (j, i).*

Note that the directions of the arcs are defined in an unconventional way. This is inspired by applications.

Definition 5 *A matrix M is called* irreducible *if its corresponding graph is strongly connected, i.e., from any node in the graph a directed path exists to any other node.*

Definition 6 *An irreducible matrix M with eigenvalue λ is order-d periodic if a k_0 exists such that $M_\otimes^{k+d} = \lambda_\otimes^d \otimes M_\otimes^k$ for all $k \geq k_0$.*

The problem of determining the order of periodicity of matrix M is equivalent to the problem of finding the most important circuit in the corresponding graph $\mathcal{G}_M$.

Definition 7 *The* average weight *of a circuit in a graph $\mathcal{G}_M$ is the sum of the weights of the individual arcs, divided by the number of arcs.*

Definition 8 *A* critical circuit *is a circuit with maximum average weight.*

The periodicity result from [4] can now be stated.

Theorem 1 *If M is irreducible and $\mathcal{G}_M$ has a unique critical circuit of length d, then M is order-d periodic with eigenvalue λ, where λ is the average weight of the critical circuit of $\mathcal{G}_M$.*

For a corresponding result in the case of non-unique critical circuits we refer to [4].

The problem of determining the order of periodicity according to Definition 6 will be called the *classical asymptoticity problem*. In this paper we are dealing with the determination of the behaviour of the characterization of the matrices U_k of Definition 9 for large k. This problem will be called the *generalized asymptoticity problem*.

Example 2 *Consider again the description of Example 1. Now suppose that the system is working gradually slower or faster in time, at a certain rate c, in such a way that the time between the kth and $(k+1)^{\text{st}}$ activities are the original times, multiplied by c^k. Then the new system description becomes*

$$x_i(k+1) = \max_{1 \leq j \leq n}(x_j(k) + c^k m_{ij}(k)). \tag{6}$$

If $m_{ij}(k) = M_{ij}$ then this can be written in max-algebra notation as

$$x(k+1) = c^k M \otimes x(k). \tag{7}$$

Systems like (7) have been studied in [1] and [5] from a shortest path point of view. The asymptotic behaviour has, however, not been considered in [1] or [5].

The generalized asymptoticity problem is concerned with the matrices U_k of the following definition. A first conjecture regarding this problem has been made in [6].

Definition 9 *For a square matrix M and a scalar $c > 0$, the matrix U_k is defined by*

$$U_k = c^{k-1} M \otimes c^{k-2} M \otimes \cdots cM \otimes M. \tag{8}$$

The reason for introducing U_k is of course that from (7) we have $x(k) = U_k \otimes x(0)$. An expression for the entries of U_k is given by

$$(U_k)_{ij} = \max_{T_k}(c^{k-1}m_{ii_1} + c^{k-2}m_{i_1i_2} + \cdots + cm_{i_{k-2}i_{k-1}} + m_{i_{k-1}j}), \qquad (9)$$

where $T_k = \{(i_1, \ldots, i_{k-1}) | i_l \in \{1, \ldots, n\}, l = 1, \ldots, k-1\}$. So every component of U_k is the maximum of n^{k-1} terms. Note that each term can be related to a path with generalized weight from j to i in the graph corresponding to M.

The rest of this paper is outlined as follows. In section 2 a fundamental relation between the cases $c < 1$ and $c > 1$ is stated; this relation allows a restriction to either $c \leq 1$ or $c \geq 1$. Section 3 contains the statement of the main theorem concerning the limiting behaviour of U_k. It is shown in section 4 that only a restricted number of cases need to be considered. The proof of the main theorem is provided in section 5. In section 6 it is argued that the generalized problem exhibits some aspects which are not present in the classical problem, viz. the influence of multiple circuits with different weights. A possible extension to larger matrices is given in section 7. Finally, section 8 states some conclusions.

2. RELATION BETWEEN RESULTS FOR $c < 1$ AND $c > 1$

We will show in this section that the generalized problem can be restricted to that of values of $c \geq 1$. A lemma which states a fundamental relationship between the cases $c < 1$ and $c > 1$ is proved.

First note that, like in linear algebra, for matrices A and B of appropriate sizes it holds that $(A \otimes B)^T = B^T \otimes A^T$. Now we can state the following lemma for general square matrices. We will use the notation $U_k(c, M)$ instead of U_k to express the dependence of U_k of the scalar c and the matrix M.

Lemma 1 *For all square matrices M and all $c > 0$,*

$$U_k(c, M) = (U_k(\frac{1}{c}, c^{k-1}M^T))^T.$$

Proof:

$$\begin{aligned}
U_k(c, M) &= c^{k-1}M \otimes c^{k-2}M \otimes \cdots \otimes cM \otimes M \\
&= (M^T \otimes cM^T \otimes \cdots \otimes c^{k-2}M^T \otimes c^{k-1}M^T)^T \\
&= (\frac{1}{c^{k-1}}(c^{k-1}M^T) \otimes \frac{1}{c^{k-2}}(c^{k-1}M^T) \otimes \cdots \otimes \frac{1}{c}(c^{k-1}M^T) \otimes (c^{k-1}M^T))^T \\
&= (U_k(\frac{1}{c}, c^{k-1}M^T))^T. \qquad (10)
\end{aligned}$$

■

Therefore, if we have determined for all M the limiting behaviour of U_k in the case $c \geq 1$, the limiting behaviour of U_k in the case $c < 1$ can be deduced from the above lemma.

3.   THE MAIN THEOREM

In this section we state the main result about the limiting behaviour of U_k in the case that M is a 2×2-matrix. We come back to general $n \times n$-matrices in section 7.

Theorem 2 Let $M = \begin{bmatrix} m_{11} & m_{12} \\ m_{21} & m_{22} \end{bmatrix}$ and $c \geq 1$. Let U_k be defined by (8).

(i) If $(cm_{12} + m_{21})/(c+1) > m_{11}$ and $(cm_{21} + m_{12})/(c+1) > m_{22}$, then for all $k \geq 1$

$$U_{2k} = U_2 + S_k, \tag{11}$$
$$U_{2k+1} = U_3 + cS_k, \tag{12}$$

where

$$S_k = \sum_{s=1}^{k-1} c^{2s} \begin{bmatrix} cm_{12} + m_{21} & cm_{12} + m_{21} \\ cm_{21} + m_{12} & cm_{21} + m_{12} \end{bmatrix}. \tag{13}$$

(ii) If $(cm_{12} + m_{21})/(c+1) \leq m_{11}$ or $(cm_{21} + m_{12})/(c+1) \leq m_{22}$, then a number k_0 and a matrix N exist such that for all $k \geq k_0$

$$U_{k+1} = cU_k + N. \tag{14}$$

As a consequence of Lemma 1 the result in the case $c < 1$ follows.

Corollary 1 Let M and U_k be defined as above, and let $c < 1$.

(i) If $(cm_{12} + m_{21})/(c+1) > m_{11}$ and $(cm_{21} + m_{12})/(c+1) > m_{22}$, then $\forall k \geq 1$, with $\bar{S}_k = (1)/(c^2)S_k^T$

$$U_{2k} = c^{2k-1}U_2 + \bar{S}_k, \tag{15}$$
$$U_{2k+1} = c^{2k}U_3 + \bar{S}_k. \tag{16}$$

(ii) If $(cm_{12} + m_{21})/(c+1) \leq m_{11}$ or $(cm_{21} + m_{12})/(c+1) \leq m_{22}$, then a number k_0 and a matrix N exist such that for all $k \geq k_0$

$$U_{k+1} = cU_k + N. \tag{17}$$

Corollary 2 In case (i), if $c > 1$ then

$$U_{k+2} - U_k = c^k \begin{bmatrix} cm_{12} + m_{21} & cm_{12} + m_{21} \\ cm_{21} + m_{12} & cm_{21} + m_{12} \end{bmatrix}. \tag{18}$$

If $c < 1$ then for case (i)

$$U_{k+2} - c^2 U_k = \begin{bmatrix} cm_{12} + m_{21} & cm_{21} + m_{12} \\ cm_{12} + m_{21} & cm_{21} + m_{12} \end{bmatrix}. \tag{19}$$

Remark 1 *Case* (i) *resembles the order-2 periodicity in the classical asymptotic-ity problem, while case* (ii) *looks like the usual order-1 periodicity. The quantities m_{11} and m_{22} are the weights of circuits of length 1; $(cm_{12} + m_{21})/(c + 1)$ and $(cm_{21} + m_{12})/(c+1)$ are the equivalents of average weights of circuits of length 2.*

The matrix U_2, which will be important for the proof of Theorem 2, can be written as

$$U_2 = \begin{bmatrix} cm_{1h} + m_{h1} & cm_{1i} + m_{i2} \\ cm_{2j} + m_{j1} & cm_{2l} + m_{l2} \end{bmatrix}, \tag{20}$$

where $h, i, j, l \in \{1, 2\}$. For specific values of h, i, j and l, we shall use the notation (h, i, j, l) and speak about 'the case (h, i, j, l)'. Clearly, (h, i, j, l) can be 16 different four-tuples.

Remark 2 *The theorem states that the only information we need, in order to know the type of asymptotic behaviour of (U_k), are the values of h and l.*

4.   RESTRICTION OF THE NUMBER OF CASES

In this section we will show that the number of cases to be considered in the proof of Theorem 2 can be reduced from 16 to 5. This reduction is achieved by noting that two cases can be dual, and by proving that some cases do not exist.
    It is easily seen that some of the cases are related in pairs, by interchanging the ones and twos of the four-tuples, and correspondingly rearranging the rows and columns of the associated matrices. We will call these pairs dual. The concept of dualizing is different from transposing! As a matter of fact, for 2×2-matrices dualizing means interchanging both the columns and the rows.

Definition 10

a) *The* dual *of matrix* $M = \begin{bmatrix} m_{11} & m_{12} \\ m_{21} & m_{22} \end{bmatrix}$ *is the matrix* $M' = \begin{bmatrix} m_{22} & m_{21} \\ m_{12} & m_{11} \end{bmatrix}$.

b) *The* dual *of case* (h, i, j, l) *is case* $(3 - l, 3 - j, 3 - i, 3 - h)$.

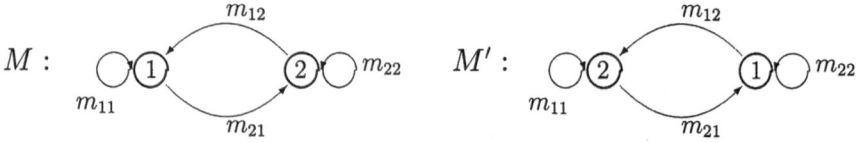

Figure 1: Two dual systems

This definition is illustrated by figure 1. By duality, if one case of each pair is analyzed, the analysis also holds for the other pair in the dual interpretation, i.e., $U_k(c, M') = U_k(c, M)'$. All pairs are given in table 1.

case	(1,1,1,1)	(1,1,1,2)	(1,1,2,1)	(1,2,1,1)	(2,1,1,1)
dual	(2,2,2,2)	(1,2,2,2)	(2,1,2,2)	(2,2,1,2)	(2,2,2,1)

case	(1,2,2,1)	(1,1,2,2)	(2,2,1,1)	(1,2,1,2)	(2,1,2,1)
dual	(2,1,1,2)	–	–	–	–

Table 1: The 16 cases restricted to 10 by duality

We will now prove that 6 of the 16 cases can not occur at all, and 2 cases can not occur for $c \geq 1$.

Lemma 2
a) The cases $(1, i, j, 1)$, $i, j \in \{1, 2\}$, reduce to the case $(1, 1, 1, 1)$.
b) The cases $(2, i, j, 2)$, $i, j \in \{1, 2\}$, reduce to the case $(2, 2, 2, 2)$.

Proof:
a) Since $h = 1$ and $l = 1$, we have

$$cm_{11} + m_{11} \geq cm_{12} + m_{21}, \tag{21}$$
$$cm_{21} + m_{12} \geq cm_{22} + m_{22}. \tag{22}$$

Suppose $i = 2$, i.e.,

$$cm_{12} + m_{22} > cm_{11} + m_{12}. \tag{23}$$

Adding (21) and (23) yields $m_{11} + m_{22} > m_{21} + m_{12}$. On the other hand, (22) + (23) yields $m_{22} + m_{11} < m_{21} + m_{12}$, which is a contradiction. A similar contradiction can be established if we suppose $j = 2$. Hence $i = 1$ and $j = 1$. Part b) follows, by duality, from part a). ∎

Finally we show that the two cases (1,2,1,2) and (2,1,2,1) can not occur for $c \geq 1$. Case (1,2,1,2):

$$cm_{11} + m_{11} \geq cm_{12} + m_{21}, \tag{24}$$
$$cm_{12} + m_{22} \geq cm_{11} + m_{12}, \tag{25}$$
$$cm_{21} + m_{11} \geq cm_{22} + m_{21}, \tag{26}$$
$$cm_{22} + m_{22} \geq cm_{21} + m_{12}. \tag{27}$$

From (24) and (25) it follows that

$$m_{11} + m_{22} \geq m_{12} + m_{21}. \tag{28}$$

Equations (25) and (26) together give

$$(1 - c)m_{11} + (1 - c)m_{22} \geq (1 - c)m_{12} + (1 - c)m_{21}, \tag{29}$$

which is compatible with (28) only if $c \leq 1$. For $c = 1$ this case reduces to (1,1,1,2), since from (25) and (26) it follows that $m_{11} = m_{22}$, so these two equations hold with equality. Hence this case can only occur for $c < 1$.

The proof that case (2,1,2,1) is only possible for $c < 1$ is similar.

Remark 3 *Noting that* $(1, 2, 1, 2)$ *and* $(1, 1, 2, 2)$ *are related by transposition (similarly for* $(2, 1, 2, 1)$ *and* $(2, 2, 1, 1)$*), it follows from the above and Lemma 1 that both cases* $(1, 1, 2, 2)$ *and* $(2, 2, 1, 1)$ *are only possible for* $c > 1$*. The two cases* $(1, 1, 2, 2)$ *and* $(1, 2, 1, 2)$ *will turn out to be key issues in the comparison with the classical problem.*

All possible cases are given in table 2.

case	(1,1,1,1)	(1,1,1,2)	(2,1,1,1)	(1,1,2,2)	(2,2,1,1)
dual	(2,2,2,2)	(1,2,2,2)	(2,2,2,1)	–	–
restriction	$c \geq 1$	$c \geq 1$	$c \geq 1$	$c > 1$	$c > 1$

Table 2: The 5 remaining cases

5. PROOF OF THE MAIN THEOREM

Proof: We will start with the proof of part (i) of the theorem. The inequalities $(cm_{12} + m_{21})/(c + 1) > m_{11}$ and $(cm_{21} + m_{12})/(c + 1) > m_{22}$ imply that we are in the case $(2, \cdot, \cdot, 1)$. From table 2 it can be concluded that only case $(2, i, 1, 1)$, $i = 1, 2$, has to be considered. So

$$U_2 = \begin{bmatrix} cm_{12} + m_{21} & cm_{1i} + m_{i2} \\ cm_{21} + m_{11} & cm_{21} + m_{12} \end{bmatrix}. \tag{30}$$

Furthermore

$$U_3 = \begin{bmatrix} (c^2 m_{11} + cm_{12} + m_{21}) & (c^2 m_{11} + cm_{1i} + m_{i2}) \\ \oplus (c^2 m_{12} + cm_{21} + m_{11}) & \oplus (c^2 m_{12} + cm_{21} + m_{12}) \\ & \\ (c^2 m_{21} + cm_{12} + m_{21}) & (c^2 m_{21} + cm_{1i} + m_{i2}) \\ \oplus (c^2 m_{22} + cm_{21} + m_{11}) & \oplus (c^2 m_{22} + cm_{21} + m_{12}) \end{bmatrix}$$
$$= \begin{bmatrix} c^2 m_{12} + cm_{21} + m_{11} & c^2 m_{12} + cm_{21} + m_{12} \\ c^2 m_{21} + cm_{12} + m_{21} & c^2 m_{21} + cm_{11} + m_{12} \end{bmatrix}. \tag{31}$$

Here $(U_3)_{21}$ is found by definitions of h and j, $(U_3)_{22}$ by definitions of i and j, $(U_3)_{12}$ by definition of h if $i = 1$, and by definitions of l and i if $i = 2$. Finally $(U_3)_{11}$ follows from inequality (32), which is a key inequality for the proof. From $cm_{12} + m_{21} \geq cm_{11} + m_{11}$ we obtain, by multiplying with $(c - 1)$, that $c^2 m_{12} - cm_{12} + cm_{21} - m_{21} \geq c^2 m_{11} - m_{11}$, and hence that

$$c^2 m_{12} + cm_{21} + m_{11} \geq c^2 m_{11} + cm_{12} + m_{21}. \tag{32}$$

We shall now prove by induction that for all $k \geq 1$

$$U_{2k} = U_2 + S_k, \tag{33}$$
$$U_{2k+1} = U_3 + cS_k. \tag{34}$$

Clearly (33) and (34) are valid for $k = 1$. Suppose that (33) and (34) hold for certain k. In the sequel we shall calculate U_{2k+2} and U_{2k+3}, both componentwise.

$$
\begin{aligned}
(U_{2k+2})_{11} &= (c^{2k+1}m_{11} + \sum_{s=1}^{k}(c^{2s}m_{12} + c^{2s-1}m_{21}) + m_{11}) \\
&\oplus(c^{2k+1}m_{12} + \sum_{s=1}^{k}(c^{2s}m_{21} + c^{2s-1}m_{12}) + m_{21}).
\end{aligned}
\tag{35}
$$

By definition of i and k applications of (32) we get

$$
c^{2k+1}m_{12} + \sum_{s=1}^{k}(c^{2s}m_{21} + c^{2s-1}m_{12}) + m_{21}
$$

$$
\geq c^{2k+1}m_{12} + \sum_{s=2}^{k}(c^{2s}m_{21} + c^{2s-1}m_{12}) + c^2 m_{21} + cm_{11} + m_{11}
$$

$$
\geq c^{2k+1}m_{12} + \sum_{s=3}^{k}(c^{2s}m_{21} + c^{2s-1}m_{12}) + c^4 m_{21} + c^3 m_{11} + c^2 m_{12} + cm_{21} + m_{11}
$$

$$
\geq \ldots
$$

$$
\geq c^{2k+1}m_{11} + \sum_{s=1}^{k}(c^{2s}m_{12} + c^{2s-1}m_{21}) + m_{11}.
\tag{36}
$$

Hence

$$
(U_{2k+2})_{11} = c^{2k+1}m_{12} + \sum_{s=1}^{k}(c^{2s}m_{21} + c^{2s-1}m_{12}) + m_{21} = (U_2)_{11} + (S_{k+1})_{11}.
\tag{37}
$$

Similarly by definition of i and k applications of (32) we get

$$
\begin{aligned}
(U_{2k+2})_{12} &= (c^{2k+1}m_{11} + \sum_{s=1}^{k}(c^{2s}m_{12} + c^{2s-1}m_{21}) + m_{12}) \\
&\oplus(c^{2k+1}m_{12} + \sum_{s=2}^{k}(c^{2s}m_{21} + c^{2s-1}m_{12}) + c^2 m_{21} + cm_{1i} + m_{i2}) \\
&= c^{2k+1}m_{12} + \sum_{s=2}^{k}(c^{2s}m_{21} + c^{2s-1}m_{12}) + c^2 m_{21} + cm_{1i} + m_{i2} \\
&= (U_2)_{12} + (S_{k+1})_{12}.
\end{aligned}
\tag{38}
$$

By k applications of (32) and by definition of j,

$$
(U_{2k+2})_{21} = c^{2k+1}m_{21} + \sum_{s=1}^{k}(c^{2s}m_{12} + c^{2s-1}m_{21}) + m_{11} = (U_2)_{21} + (S_{k+1})_{21},
\tag{39}
$$

and finally by definition of $(U_3)_{12}$, $(k-1)$ applications of (32) and by definition of j,

$$
(U_{2k+2})_{22} = c^{2k+1}m_{21} + \sum_{s=1}^{k}(c^{2s}m_{12} + c^{2s-1}m_{21}) + m_{12} = (U_2)_{22} + (S_{k+1})_{22}.
\tag{40}
$$

Hence U_{2k+2} satisfies (33).

Next we use U_{2k+2} to calculate U_{2k+3}. Using (32) $(k+1)$ times we find

$$
\begin{aligned}
(U_{2k+3})_{11} &= (c^{2k+2}m_{11} + \sum_{s=0}^{k}(c^{2s+1}m_{12} + c^{2s}m_{21})) \\
&\oplus(c^{2k+2}m_{12} + \sum_{s=1}^{k}(c^{2s+1}m_{21} + c^{2s}m_{12}) + cm_{21} + m_{11}) \\
&= c^{2k+2}m_{12} + \sum_{s=1}^{k}(c^{2s+1}m_{21} + c^{2s}m_{12}) + cm_{21} + m_{11} \\
&= (U_3)_{11} + c(S_{k+1})_{11}.
\end{aligned} \tag{41}
$$

By definition we have

$$
\begin{aligned}
(U_{2k+3})_{12} &= (c^{2k+2}m_{11} + \sum_{s=1}^{k}(c^{2s+1}m_{12} + c^{2s}m_{21}) + cm_{1i} + m_{i2}) \\
&\oplus(c^{2k+2}m_{12} + \sum_{s=0}^{k}(c^{2s+1}m_{21} + c^{2s}m_{12})).
\end{aligned} \tag{42}
$$

By k applications of (32) we find

$$
\begin{aligned}
&c^{2k+2}m_{11} + \sum_{s=1}^{k}(c^{2s+1}m_{12} + c^{2s}m_{21}) + cm_{1i} + m_{i2} \\
&\leq c^{2k+2}m_{12} + \sum_{s=2}^{k}(c^{2s+1}m_{21} + c^{2s}m_{12}) + c^3m_{21} + c^2m_{11} + cm_{1i} + m_{i2}.
\end{aligned} \tag{43}
$$

Furthermore, by definition of h if $i = 1$, and by definitions of i and l if $i = 2$, we have

$$
c^2m_{11} + cm_{1i} + m_{i2} \leq c^2m_{12} + cm_{21} + m_{12}. \tag{44}
$$

Hence we conclude

$$
(U_{2k+3})_{12} = c^{2k+2}m_{12} + \sum_{s=0}^{k}(c^{2s+1}m_{21} + c^{2s}m_{12}) = (U_3)_{12} + c(S_{k+1})_{12}. \tag{45}
$$

By definition of j, k applications of (32) and by definition of i, we have

$$
\begin{aligned}
(U_{2k+3})_{21} &= (c^{2k+2}m_{21} + \sum_{s=0}^{k}(c^{2s+1}m_{12} + c^{2s}m_{21})) \\
&\oplus(c^{2k+2}m_{22} + \sum_{s=1}^{k}(c^{2s+1}m_{21} + c^{2s}m_{12}) + cm_{21} + m_{11}) \\
&= c^{2k+2}m_{21} + \sum_{s=0}^{k}(c^{2s+1}m_{12} + c^{2s}m_{21}) = (U_3)_{21} + c(S_{k+1})_{21}.
\end{aligned} \tag{46}
$$

Finally by definition of j, k applications of (32) and by definition of i,

$$
\begin{aligned}
(U_{2k+3})_{22} &= \left(c^{2k+2}m_{21} + \sum_{s=1}^{k}(c^{2s+1}m_{12} + c^{2s}m_{21}) + cm_{1i} + m_{i2}\right) \\
&\quad \oplus \left(c^{2k+2}m_{22} + \sum_{s=0}^{k}(c^{2s+1}m_{21} + c^{2s}m_{12})\right) \\
&= c^{2k+2}m_{21} + \sum_{s=1}^{k}(c^{2s+1}m_{12} + c^{2s}m_{21}) + cm_{1i} + m_{i2} \\
&= (U_3)_{22} + c(S_{k+1})_{22}.
\end{aligned}
\tag{47}
$$

This concludes the proof of case (i).

For case (ii), in all cases except (1,1,1,2) and (1,2,2,2) general expressions for U_k can be easily derived by straightforward induction, see table 3. From these expressions the result of the theorem follows, with $k_0 = 2$. In the cases (1,1,1,2) and (1,2,2,2), however, k_0 may reach an arbitrarily large value, and for the proof some more effort is needed. The cases are dual, so let us only consider the first one. Like in the previous cases, it is not difficult to prove by induction that U_k has the general expression

$$
\begin{bmatrix}
\displaystyle\sum_{s=0}^{k-1}c^s m_{11} & \displaystyle\sum_{s=1}^{k-1}c^s m_{11} + m_{12} \\
c^{k-1}m_{21} + \displaystyle\sum_{s=0}^{k-2}c^s m_{11} & \left(c^{k-1}m_{21} + \displaystyle\sum_{s=1}^{k-2}c^s m_{11} + m_{12}\right) \oplus \left(\displaystyle\sum_{s=0}^{k-1}m_{22}\right)
\end{bmatrix}.
\tag{48}
$$

We have the following inequalities:

$$
\begin{aligned}
cm_{11} + m_{11} &\geq cm_{12} + m_{21}, & (49) \\
cm_{11} + m_{12} &\geq cm_{12} + m_{22}, & (50) \\
cm_{21} + m_{11} &\geq cm_{22} + m_{21}, & (51) \\
cm_{22} + m_{22} &\geq cm_{21} + m_{12}. & (52)
\end{aligned}
$$

First the case $c = 1$ is dealt with. In this case

$$
\begin{aligned}
2m_{11} &\geq m_{12} + m_{21}, & (53) \\
2m_{22} &\geq m_{12} + m_{21}, & (54) \\
m_{11} &> m_{22}. & (55)
\end{aligned}
$$

Let us suppose that $m_{11} = m_{22} + \gamma_1$, $m_{21} + m_{12} = 2m_{22} - \gamma_2$, where $\gamma_1 > 0, \gamma_2 \geq 0$. Then

$$
m_{21} + (k-2)m_{11} + m_{12} = km_{22} + (k-2)\gamma_1 - \gamma_2 \geq km_{22}
\tag{56}
$$

for k large enough, viz.

$$
k \geq \xi_0 \equiv \frac{\gamma_2 + 2\gamma_1}{\gamma_1} = \frac{2m_{11} - m_{21} - m_{12}}{m_{11} - m_{22}}.
\tag{57}
$$

This proves (58) in case $c = 1$, where $k_0 \geq \xi_0$ (because $\xi_0 \in \mathbb{R}$ and $k_0 \in \mathbb{N}$).

We will have to use (49)-(52) in a more sophisticated way than we did previously, in order to be able to prove that a k_0 exists such that

$$(U_k)_{22} = c^{k-1}m_{21} + \sum_{s=1}^{k-2} c^s m_{11} + m_{12} \qquad \text{for all } k \geq k_0. \qquad (58)$$

More specifically, the inequalities in (50) and (51) can be supposed to be strict, since otherwise this case reduces to one which has already been treated. Therefore we obtain

$$m_{22} < cm_{11} + (1-c)m_{12}, \qquad (59)$$

$$m_{22} < \frac{1}{c}m_{11} + \frac{c-1}{c}m_{21}. \qquad (60)$$

Note the impact of Lemma 1 here: replacing c by $\frac{1}{c}$ and m_{12} by m_{21} in (59) yields (60).

Let $c > 1$. An expression for k_0 can be found by considering that for $k \geq k_0$ it must hold that $U_{k+1} - cU_k = N$, hence k_0 is the smallest value of k for which it holds that

$$\left(c^{k-1}m_{21} + \sum_{s=1}^{k-2} c^s m_{11} + m_{12}\right) \geq \left(\sum_{s=0}^{k-1} m_{22}\right)$$

$$\Leftrightarrow \quad c^{k-1}m_{21} + c\frac{1-c^{k-2}}{1-c}m_{11} + m_{12} \geq \frac{1-c^k}{1-c}m_{22}$$

$$\Leftrightarrow \quad c^{k-1}\left(m_{21} - \frac{1}{1-c}m_{11} + \frac{c}{1-c}m_{22}\right) \geq \frac{1}{1-c}m_{22} - \frac{c}{1-c}m_{11} - m_{12}$$

$$\Leftrightarrow \quad c^{k-1} \geq \frac{m_{22} - cm_{11} - (1-c)m_{12}}{cm_{22} - m_{11} + (1-c)m_{21}}$$

$$\Leftrightarrow \quad k \geq \log\left(\frac{m_{22} - cm_{11} - (1-c)m_{12}}{cm_{22} - m_{11} + (1-c)m_{21}}\right)/\log(c) + 1 \equiv \xi_0. \qquad (61)$$

This completes the proof of (ii). ∎

For the case $c < 1$, (61) can be proved similarly, but if we apply Lemma 1 then it follows from (61) (interchange m_{21} and m_{12}, replace c with $\frac{1}{c}$) that k_0 is the smallest value for k such that

$$k \geq \log\left(\frac{m_{22} - \frac{1}{c}m_{11} - (1-\frac{1}{c})m_{21}}{\frac{1}{c}m_{22} - m_{11} + (1-\frac{1}{c})m_{12}}\right)/\log(\frac{1}{c}) + 1$$

$$= \log\left(\frac{cm_{22} - m_{11} - (c-1)m_{21}}{m_{22} - cm_{11} + (c-1)m_{12}}\right)/\log(\frac{1}{c}) + 1 \quad = \xi_0. \qquad (62)$$

We could of course have given this result immediately applying Lemma 1, but the above calculation illustrates the validity of the expression for ξ_0.

6.   CLASSICAL VERSUS GENERALIZED PROBLEM

In the classical problem an irreducible matrix has a well-understood asymptotic behavior. The order of periodicity is the length of the critical circuit, and the unique eigenvalue determines the asymptotic speed. If several circuits are critical, the least common multiple of the lengths of these circuits is the order of periodicity, but still the unique eigenvalue is the rate at which the entire system operates.

In the generalized problem, though, we showed that it is possible that multiple rates can determine the system's asymptotic behavior. From table 3 we see that in case $(1,1,2,2)$ (and so in case $(1,2,1,2)$) both m_{11} and m_{22} are part of the asymptotic behavior. This feature of several circuits remaining present asymptotically seems to be characteristic for the generalized problem for general square matrices. In the following section we elaborate on this. The question *which* circuits are the ones that determine the asymptoticity remains as yet unanswered, but a selection of candidates is made on the basis of the idea that a circuit can only be critical if for each node in the circuit it has the largest generalized weight. This idea is expressed by formula (64) which collects all these preselected circuits in a set Γ.

7.   EXTENSIONS TO HIGHER DIMENSIONS

In the case of a two-dimensional system it is observed that the matrix is order-d periodic if the generalized critical circuit is of length d, $d = 1, 2$. A possible extension can be that for a general matrix the presence of a unique generalized critical circuit of length d implies generalized order-d periodicity. Apparently with each node a critical circuit can be associated. Define the generalized weight of a circuit $\gamma = i \rightarrow i_1 \rightarrow \cdots i_{d-1} \rightarrow i$ of length d, starting from node $i \in \gamma$, as

$$W(\gamma, i) \equiv \frac{m_{i_1 i} + c m_{i_2 i_1} + \cdots + c^{d-2} m_{i_{d-1} i_{d-2}} + c^{d-1} m_{i i_{d-1}}}{1 + c + \cdots + c^{d-2} + c^{d-1}}. \tag{63}$$

By a *critical circuit with respect to node* i_0 in the generalized setup we mean a circuit γ_{i_0} containing node i_0 for which $W(\gamma_{i_0}, i_0) \geq W(\gamma, i_0)$ for all other circuits γ containing i_0. So $W(\gamma_{i_0}, i_0) = max_{\gamma \ni i_0} W(\gamma, i_0)$. In this way a circuit γ_i is associated with each node i. Next define the set Γ containing some of these circuits γ_i as

$$\Gamma = \{\gamma_i | \gamma_i = \gamma_j \text{ for all } j \in \gamma_i\}. \tag{64}$$

All circuits in Γ will be called *generalized critical circuits*. So any generalized critical circuit is critical with respect to *all* its nodes.

With this definition of the generalized critical circuits, we state the following extension if M is an $n \times n$-matrix. The extension is inspired by numerical experiments. Current research focuses on results from other areas of mathematics with which it can be proved.

We have $\Gamma = \{\gamma_{i_1}, \ldots, \gamma_{i_l}\}$ for some l. Then M is order-r periodic, where r is the least common multiple of the lengths of *some* of the generalized critical circuits. As noted in the preceding section it is not exactly clear which circuits these should be, but it is clear that these circuits are such that they do not influence each other as k tends to infinity. This is what happens in the case studied in this paper.

A special case occurs when there is exactly one generalized critical circuit, i.e., when $\Gamma = \{\gamma_i\}$ (some nodes may not be included in this circuit). Then clearly M is generalized order-r periodic, where r is the length of this circuit γ_i.

An even more restricted case is that with all nodes $1, \ldots, n$ the same critical circuit is associated, i.e., $\Gamma = \{\gamma\}$ where $\gamma = \gamma_1 = \gamma_2 = \ldots = \gamma_n$. From the above it follows that M is generalized order-n periodic.

8. CONCLUSIONS

In this article we stated a theorem concerning the asymptotic behavior of a generalized class of systems, e.g. systems that are speeding up or slowing down with a certain rate c. In the usual max algebra, in the case of an irreducible matrix, the periodic behavior is determined by one circuit, viz. the critical circuit, and the corresponding eigenvalue. In the generalized problem discussed in this paper, however, it is possible that two circuits influence the asymptotic system. This is seen in the cases $(1,1,2,2)$ and $(1,2,1,2)$.

Some numerical experiments give rise to a formulation of an extension to higher dimensions. In this general framework some results have already been obtained and it is expected that the full problem treatment will soon be published.

REFERENCES

[1] M. Gondran and M. Minoux. *Graphs and algorithms*. Wiley, 1984.

[2] R.A. Cuninghame-Green. *Minimax algebra*. Springer-Verlag, New York, U.S.A., 1979.

[3] U. Zimmermann. *Linear and combinatorial optimization in ordered algebraic structures*, volume 10 of *Annals of discrete mathematics*. North-Holland, 1981.

[4] G. Cohen, D. Dubois, J.P. Quadrat, and M. Viot. A linear-system-theoretic view of discrete-event processes and its use for performance evaluation in manufacturing. *IEEE Transactions on Automatic Control*, 30:210–220, 1985.

[5] M. Minoux. Structures algébriques généralisées des problèmes de cheminement dans les graphes. *RAIRO Recherche opérationnelle*, 10:33–62, 1976.

[6] J.A.C. Resing. *Asymptotic results in feedback systems*. PhD thesis, Delft University of Technology, Holland, 1990.

(h,i,j,l)	k_0	U_k
$(1,1,1,1)$	2	$\begin{bmatrix} \sum_{s=0}^{k-1} c^s m_{11} & \sum_{s=1}^{k-1} c^s m_{11} + m_{12} \\ c^{k-1} m_{21} + \sum_{s=0}^{k-2} c^s m_{11} & c^{k-1} m_{21} + \sum_{s=1}^{k-2} c^s m_{11} + m_{12} \end{bmatrix}$
$(2,2,2,2)$	2	$\begin{bmatrix} c^{k-1} m_{12} + \sum_{s=1}^{k-2} c^s m_{22} + m_{21} & c^{k-1} m_{12} + \sum_{s=0}^{k-2} c^s m_{22} \\ \sum_{s=1}^{k-1} c^s m_{22} + m_{21} & \sum_{s=0}^{k-1} c^s m_{22} \end{bmatrix}$
$(1,1,1,2)$	ξ_0	$k \geq k_0:$ $\begin{bmatrix} \sum_{s=0}^{k-1} c^s m_{11} & \sum_{s=1}^{k-1} c^s m_{11} + m_{12} \\ c^{k-1} m_{21} + \sum_{s=0}^{k-2} c^s m_{11} & c^{k-1} m_{21} + \sum_{s=1}^{k-2} c^s m_{11} + m_{12} \end{bmatrix}$ $k < k_0:$ $\begin{bmatrix} \sum_{s=0}^{k-1} c^s m_{11} & \sum_{s=1}^{k-1} c^s m_{11} + m_{12} \\ c^{k-1} m_{21} + \sum_{s=0}^{k-2} c^s m_{11} & \sum_{s=0}^{k-1} c^s m_{22} \end{bmatrix}$
$(1,2,2,2)$	ξ_0	$k \geq k_0:$ $\begin{bmatrix} c^{k-1} m_{12} + \sum_{s=1}^{k-2} c^s m_{22} + m_{21} & c^{k-1} m_{12} + \sum_{s=0}^{k-2} c^s m_{22} \\ \sum_{s=1}^{k-1} c^s m_{22} + m_{21} & \sum_{s=0}^{k-1} c^s m_{22} \end{bmatrix}$ $k < k_0:$ $\begin{bmatrix} \sum_{s=0}^{k-1} c^s m_{11} & c^{k-1} m_{12} + \sum_{s=0}^{k-2} c^s m_{22} \\ \sum_{s=1}^{k-1} c^s m_{22} + m_{21} & \sum_{s=0}^{k-1} c^s m_{22} \end{bmatrix}$
$(1,1,2,2)$ $c > 1$	2	$\begin{bmatrix} \sum_{s=0}^{k-1} c^s m_{11} & \sum_{s=1}^{k-1} c^s m_{11} + m_{12} \\ \sum_{s=1}^{k-1} c^s m_{22} + m_{21} & \sum_{s=0}^{k-1} c^s m_{22} \end{bmatrix}$

Table 3: U_k in case (ii)

CONDITIONS FOR TRACKING TIMING PERTURBATIONS IN TIMED PETRI NETS WITH MONITORS*

Kyle J. Williams   Julie A. Gannon   Mark S. Andersland

James E. Lumpp, Jr.   Thomas L. Casavant [†]

Abstract. This work concerns the tracking of timing perturbations in discrete event dynamic systems (DEDS) modeled by timed Petri nets (TPN). Necessary and sufficient conditions are presented for tracking perturbations in TPNs in which timings are only available for a set of monitor transitions. These conditions imply that tracking can be performed in some TPNs without knowledge of all timings. We present a class of safe TPNs as an example.

1. INTRODUCTION

This work concerns the tracking of timing perturbations in discrete event dynamic systems (DEDS) modeled by timed Petri nets (TPN). Analysis of this sort was first considered by Ho et. al. [1, 2] in the context of queueing network simulation and optimization. Our interest is motivated by the problem of accounting for the intrusion that arises when software probes (e.g., print statements) are used to monitor the internal state of executing, asynchronous, distributed software. When a software probe delays one process relative to another, the delay may change program behavior to the point that the observations become invalid. Our idea [3] is to compensate for the monitoring intrusion by tracking, in a TPN model of the monitored software [4], the timing perturbations that would be induced if the software probe delays were set to zero.

An algorithm that tracks timing perturbations in TPNs is described in [5]. In this paper we consider the problem of determining what timing information is necessary and sufficient to track timing perturbations in TPNs given the firing times of a set of monitor transitions. To identify necessary and sufficient conditions for tracking, one must first specify what a priori timing information is available. We assume that only the monitor transition firing times are known, and that only the monitor transition firing delays are subject to perturbation. Under these assumptions, we identify the information necessary and sufficient for tracking timing perturbations and consider methods for acquiring this information. We demonstrate that, for a class of safe TPNs, tracking perturbations is possible without knowing all transition firing times, because some timing information can be derived from TPN structure.

*This work supported by DARPA grant N00174-91-C-0116, NSF Grant CCR-9024484, and the Hewlett Packard Faculty Development Program.

†Electrical and Computer Engineering Department, The University of Iowa, Iowa City, Iowa 52242-1595 USA; e-mail: {kwilliam, jagannon, msanders, jel, tomc}@eng.uiowa.edu

In Section 2 we define the class of nets under consideration and provide definitions. We formulate the perturbation tracking problem in Section 3, and derive preliminary results in Section 4. In Section 5, we show that tracking perturbations is possible if and only if the induced perturbations and the *idle times* upstream from monitor transitions are known, where an *idle time* is the variable delay a token incurs at a place before enabling a transition. If this information can not be derived from structure, it must be determined from the monitor transition firing times. In Section 6, we demonstrate how structure can be combined with observed firing times to provide the necessary and sufficient information for a class of safe TPNs.

2. NOTATION

For the purposes of our study, a TPN is a bipartite, directed graph of transitions, places, and arcs, denoted by the 4-tuple $\mathcal{P} := (T, P, A, x_0)$, where

T is a finite set of transitions;

P is a finite set of places;

$A \subset (P \times T) \cup (T \times P)$ is a set of directed arcs;

x_0 is a function $x_0 : P \rightarrow \{0, 1, \ldots\}$, returning the initial number of tokens in place p.

The TPN is ordinary, first-in first-out (FIFO), and has places with unbounded capacity. The timing of the net is determined by the net's *firing delay schedule* $V := \{v_t : t \in T\}$, $v_t = \{v_t(1), v_t(2), \ldots\}$. Here $v_t(k) \in \mathbb{R}^+$, $k \in \mathbb{N}$, denotes the firing delay incurred by the kth firing of transition t.[1] The normal firing rules apply [7]. Although V may be random, the behavior of the net is assumed deterministic for fixed V. In particular, this implies that the behavior of tokens at decision places is deterministic.[2] For example, nets with no decision places [8] and nets with decision places that are controlled by exogenous decision sequences [7] are deterministic. In essence, fixing V imposes a unique total ordering on the firing times of transitions that interact. Hence, V completely determines the system behavior. When it is necessary to distinguish between the two behaviors, we denote the firing delay schedule in the unperturbed and perturbed nets by $\overline{V}$ and V, respectively. Similarly, quantities in the unperturbed net will be indicated by overbars.

We use, for $t \in T$ and $k \in \mathbb{N}$, the following notation:

---

[1]Note that the firing schedule V is different from the clock structure defined in [6].

[2]A decision place is a place with more than one output transition.

$\mathcal{I}(u), \mathcal{O}(u)$ are the input and output sets, respectively, of transition or place u, e.g., $\mathcal{I}(t) = \{p \in P : \exists\, (p,t) \in A\}$. The notation extends in the usual way to sets of places or transitions.

$\langle t, k \rangle$ denotes the kth firing of t.

$\tau_t(k)$ is the *firing time* of $\langle t, k \rangle$, i.e., the time that t completes firing for the kth time. τ_t is the sequence of these times for t. By convention, $\tau_t(k) = \infty$ if t never fires for the kth time.

$\Delta_t(k)$ $:= v_t(k) - \bar{v}_t(k)$, is the *perturbation* induced on the kth firing of t. Δ_t is the sequence of these perturbations for t.

$M \subset T$ is the set of *monitors*, i.e., transitions t for which $\bar{\tau}_t$ are known and $\Delta_t \neq 0$. Transitions $t \in T \setminus M$ have $\bar{\tau}_t$ unknown, $\Delta_t = 0$.

Because the behavior of the net is deterministic for fixed V, we can also define the following notation:

$C_V(\langle t, k \rangle)$ $:= \{\langle u, \ell \rangle,\ u \in \mathcal{I}(\mathcal{I}(t)),\ \ell \in \mathbb{N} : \langle u, \ell \rangle$ produces, for fixed V, a token consumed by $\langle t, k \rangle\}$. In general, $|\,C_V(\langle t, k \rangle)\,| \leq |\,\mathcal{I}(t)\,|$ because $\langle t, k \rangle$ may be consuming a token initially in some $p \in \mathcal{I}(t)$.

$I_V(\langle u, \ell \rangle; \langle t, k \rangle)$ $:= \tau_t(k) - \tau_u(\ell) - v_t(k),\ \langle u, \ell \rangle \in C_V(\langle t, k \rangle)$, is the *idle time* for fixed V that the token produced by $\langle u, \ell \rangle$ incurs at an intervening place before it enables $\langle t, k \rangle$.

Using the characterization of "upstream" provided by $C_V(\cdot)$, we make, for fixed V, the following definitions:

Definition 1 *A token track is a sequence of transition firings* $\{\langle s_i, h_i \rangle\}_{i=1}^n$, $n > 1$, *such that* $\langle s_{i-1}, h_{i-1} \rangle \in C_V(\langle s_i, h_i \rangle)$, *for* $i = 2, 3, \ldots, n$.

Definition 2 *A token track* $\{\langle s_i, h_i \rangle\}_{i=1}^n$ *is rooted when* $s_n \in M$, $s_j \notin M$ *for all* $j = 2, 3, \ldots, n - 1$, *and* $\langle s_1, h_1 \rangle$ *satisfies at least one of the following conditions:* *(a)* $s_1 \in M$, *(b)* $|\,\mathcal{I}(s_1)\,| = 0$, *or* *(c)* $|\,\mathcal{I}(s_1)\,| = 1$ *and* $x_0(\mathcal{I}(s_1)) \geq h_1$.

Token tracks are snapshots of TPN behavior analogous to the paths of the *Evolution Tree* defined in [7] for places. The firings of transitions satisfying (b) or (c) of definition 2 are termed *initial firings*.

We define $D_\mathcal{P}$ to be the set of all possible rooted token tracks that exist in TPN $\mathcal{P}$ for some V. In the worst case, we can not determine which token tracks are possible, and $D_\mathcal{P}$ contains every rooted token track. Additional structural information may eliminate token tracks from $D_\mathcal{P}$. For a fixed V, let $D_V \subseteq D_\mathcal{P}$ be the set of rooted token tracks that exist for V.

3. PROBLEM STATEMENT

To develop necessary conditions for tracking perturbations in TPNs, we make the following assumptions:

A1. Firing delays $\bar{v}_t(k)$ and $v_t(k)$, $t \in T \setminus M$, $k \in \mathbb{N}$, only depend on t's firing index k, not on t's enabling time.

A2. For all rooted token tracks $\{\langle s_i, h_i \rangle\}_{i=1}^n \in D_{\mathcal{P}}$, s_1, h_1, s_n, and h_n are known, i.e., the transition label and firing number of each track's first and last firings are known.

A3. For all rooted token tracks $S \in D_{\mathcal{P}}$, if $S \in D_{\overline{V}}$ then $S \in D_V$, i.e., rooted token tracks do not change under perturbation.

A4. For all rooted token tracks $S = \{\langle s_i, h_i \rangle\}_{i=1}^n \in D_{\mathcal{P}}$ and index set $\overline{B} \subseteq \{2, 3, \ldots, n\}$, there exists a $\overline{V}$ such that $S \in D_{\overline{V}}$, $\overline{I}_{\overline{V}}(\langle s_{j-1}, h_{j-1} \rangle; \langle s_j, h_j \rangle) = 0$ for all $j \in \overline{B}$, and $\overline{I}_{\overline{V}}(\langle s_{k-1}, h_{k-1} \rangle; \langle s_k, h_k \rangle) \neq 0$ for all $j \notin \overline{B}$.

A5. A4 holds with $\overline{B}$ and $\overline{V}$ replaced by B and V, respectively.

Assumption $A1$ ensures that the non-monitors are "time-invariant", i.e., that $\overline{v}_t(k) = v_t(k)$, for all $t \in T \setminus M$, $k \in \mathbb{N}$. This implies that non-monitors do not induce "second-order" perturbations. $A2$ ensures that we can follow perturbations through the net.

Tracking perturbations becomes more difficult when token tracks change due to timing perturbations. $A3$ ensures that, aside from changes in transition firing times, each transition firing in a token track produces and consumes the same tokens under perturbation. This assumption applies to at least one firing of all token tracks rooted by an initial firing.

A4. and A5. imply that no information concerning relative TPN timings is available. In particular, these assumptions imply that the firing delay schedule is unknown. In this sense, the assumptions bound the known timing information, and lead to the strongest necessary conditions.

The 0th order perturbation tracking problem (tracking up to a token track change) can be stated as follows.

(TP0:)   Given a TPN satisfying assumptions $A1$ - $A5$, given that the firing times $\overline{\tau}_t$ of the monitors $t \in M$ are known, and given that the firing durations $\overline{v}_t$ of these transitions are perturbed, find the sequence of perturbed firing times τ_t of the transitions $t \in M$.

4. PRELIMINARY RESULTS

Let $\{\langle s_i, h_i \rangle\}_{i=1}^n$, $n > 1$, be a token track in the unperturbed net. Because $\overline{\tau}_{s_n}(h_n)$ can be expressed as a telescoping sum along this track

$$\overline{\tau}_{s_n}(h_n) = \overline{\tau}_{s_1}(h_1) + \sum_{i=2}^n \left(\overline{\tau}_{s_i}(h_i) - \overline{\tau}_{s_{i-1}}(h_{i-1}) - \overline{v}_{s_i}(h_i) \right) + \sum_{i=2}^n \overline{v}_{s_i}(h_i)$$

$$= \overline{\tau}_{s_1}(h_1) + \sum_{i=2}^n \overline{I}_{\overline{V}}(\langle s_{i-1}, h_{i-1} \rangle; \langle s_i, h_i \rangle) + \sum_{i=2}^n \overline{v}_{s_i}(h_i). \tag{1}$$

Because $A3$ ensures that the token track is the same in the perturbed net, we can also write

$$\tau_{s_n}(h_n) = \tau_{s_1}(h_1) + \sum_{i=2}^n I_V(\langle s_{i-1}, h_{i-1} \rangle; \langle s_i, h_i \rangle) + \sum_{i=2}^n v_{s_i}(h_i). \tag{2}$$

Subtracting (2) from (1) and using the definition of $\Delta_{s_i}(h_i)$, we find that

$$
\overline{\tau}_{s_n}(h_n) - \tau_{s_n}(h_n) = \overline{\tau}_{s_1}(h_1) - \tau_{s_1}(h_1) + \sum_{i=2}^{n} \left(\overline{I}_{\overline{V}}(\langle s_{i-1}, h_{i-1} \rangle; \langle s_i, h_i \rangle) \right.
$$
$$
\left. - I_V(\langle s_{i-1}, h_{i-1} \rangle; \langle s_i, h_i \rangle) \right) - \sum_{i=2}^{n} \Delta_{s_i}(h_i). \qquad (3)
$$

When $\{\langle s_i, h_i \rangle\}_{i=1}^{n} \in D_{\mathcal{P}}$ (i.e., when $\{\langle s_i, h_i \rangle\}_{i=1}^{n}$ is a rooted token track), by definition, $\Delta_{s_i}(h_i) = 0$, for $i = 2, 3, \ldots, n-1$; hence

$$
\overline{\tau}_{s_n}(h_n) - \tau_{s_n}(h_n) = \overline{\tau}_{s_1}(h_1) - \tau_{s_1}(h_1) + \sum_{i=2}^{n} \left(\overline{I}_{\overline{V}}(\langle s_{i-1}, h_{i-1} \rangle; \langle s_i, h_i \rangle) \right.
$$
$$
\left. - I_V(\langle s_{i-1}, h_{i-1} \rangle; \langle s_i, h_i \rangle) \right) - \Delta_{s_n}(h_n). \qquad (4)
$$

To determine necessary and sufficient conditions for solving TP0 under *A1* - *A5*, we require two results.

Lemma 1 *For a fixed V, for all $\langle t, k \rangle$, with $t \in M$ and $k \in \mathbb{N}$ such that $\tau_t(k) < \infty$ and $\langle t, k \rangle$ is not an initial firing, there exists a rooted token track $\{\langle s_i, h_i \rangle\}_{i=1}^{n} \in D_V$, with $n > 1$, such that $\langle t, k \rangle \twoheadrightarrow \langle s_n, h_n \rangle$.*

Proof: Fix V and $\langle t, k \rangle$, $t \in M$ and $k \in \mathbb{N}$, such that $\tau_t(k) < \infty$ and $\langle t, k \rangle$ is not an initial firing. Because $\langle t, k \rangle$ is not an initial firing, there exists some transition firing $\langle u, \ell \rangle \in C_V(\langle t, k \rangle)$. By definition 1, $S = \{\langle u, \ell \rangle, \langle t, k \rangle\}$ is a token track under V. Clearly S has length greater than one. Let $\langle s_1, h_1 \rangle$ denote the first transition firing in S. If $\langle s_1, h_1 \rangle$ satisfies one of the three conditions of definition 2, then S is a rooted token track. If not, then choose any $\langle q, f \rangle \in C_V(\langle s_1, h_1 \rangle)$, attach it to the beginning of S, and call this new token track S. Repeat this process, of adding a transition firing to the beginning of S, until the first transition firing $\langle s_1, h_1 \rangle$ satisfies one of the conditions for the first firing of a rooted token track. Even if $\langle s_1, h_1 \rangle$ never satisfies condition (a), it must eventually satisfy either (b) or (c). Hence, $S \in D_V$ is a rooted token track of length greater than one that terminates at $\langle t, k \rangle$. ∎

Lemma 1 ensures that if a monitor firing is not an initial firing it terminates some rooted token track for every possible V. This implies that we can use rooted token tracks to characterize the behavior of the net upstream from the monitor.

Lemma 2 *Fix $\overline{V}$ and $\{\langle s_i, h_i \rangle\}_{i=1}^{n} \in D_{\overline{V}}$. Assuming that τ_t is known for all $t \in M$, $\overline{\tau}_{s_1}(h_1) - \tau_{s_1}(h_1)$ is known.*

Proof: By definition 2, $\langle s_1, h_1 \rangle$ must satisfy at least one of the three rooted token track conditions. It suffices to consider these cases separately.

Case 1: For all $h_1 \in \mathbb{N}$, when $s_1 \in M$, $\overline{\tau}_{s_1}(h_1)$ is known by definition, and $\tau_{s_1}(h_1)$ is known by the assumption of the Lemma. Hence, $\overline{\tau}_{s_1}(h_1) - \tau_{s_1}(h_1)$ is known.

Case 2: Suppose $|\mathcal{I}(s_1)| = 0$. When $s_1 \in M$, then as in Case 1, $\overline{\tau}_{s_1}(h_1) - \tau_{s_1}(h_1)$ is known. Suppose $s_1 \notin M$. Because transition s_1 is enabled for all $h_1 \in \mathbb{N}$,

$$\overline{\tau}_{s_1}(h_1) = \sum_{j=1}^{h_1} \overline{v}_{s_1}(j). \tag{5}$$

By the definition of M, in the perturbed net, $\Delta_{s_1}(h_1) = 0$ for all $h_1 \in \mathbb{N}$, and (5) becomes

$$\tau_{s_1}(h_1) = \sum_{j=1}^{h_1} v_{s_1}(j) = \sum_{j=1}^{h_1} \left(\overline{v}_{s_1}(j) + \Delta_{s_1}(j) \right) = \sum_{j=1}^{h_1} \overline{v}_{s_1}(j). \tag{6}$$

Subtracting (6) from (5) we see that $\overline{\tau}_{s_1}(h_1) - \tau_{s_1}(h_1) = 0$ is known.

Case 3: When $|\mathcal{I}(s_1)| = 1$ and $x_0(\mathcal{I}(s_1)) > h_1$, s_1 has been constantly enabled by initial tokens because the net is FIFO. Hence, as in Case 2, $\overline{\tau}_{s_1}(h_1) - \tau_{s_1}(h_1) = 0$ is known. ∎

When TP0 has been solved, τ_t is known for all $t \in M$. Accordingly, Lemma 2 implies that the cumulative perturbation $\overline{\tau}_{s_1}(h_1) - \tau_{s_1}(h_1)$ is known when TP0 has been solved. We will use this fact, along with Lemma 1, to track perturbations along rooted token tracks.

5.   NECESSARY AND SUFFICIENT CONDITIONS

Theorem 1 *To solve perturbation tracking problem TP0, it is necessary and sufficient to know*

>   *N1: the perturbation $\Delta_t(k)$, for all $t \in M$, $k \in \mathbb{N}$.*
>   *N2: the idle times $\overline{I}_{\overline{V}}(\langle s_{b-1}, h_{b-1} \rangle; \langle s_b, h_b \rangle)$, $I_V(\langle s_{b-1}, h_{b-1} \rangle; \langle s_b, h_b \rangle)$,*
>   *$b = 2, 3, \ldots, n$, for all rooted token tracks $S = \{\langle s_i, h_i \rangle\}_{i=1}^n \in D_p$.*

Proof:

Necessity: Assume that the solution to TP0 is known, i.e., $\tau_u(\ell)$ is known for all $u \in M$, $\ell \in \mathbb{N}$. Consider (4) written for any rooted token track $S = \{\langle s_i, h_i \rangle\}_{i=1}^n$. $\overline{\tau}_{s_n}(h_n) \in M$ is known by the definition of M; $\tau_{s_n}(h_n)$ is known by the hypothesis; and $\overline{\tau}_{s_1}(h_1) - \tau_{s_1}(h_1)$ is known by Lemma 2. Hence, for all rooted token tracks S, only the perturbation $\Delta_{s_n}(h_n)$, and the idle times $\overline{I}_{\overline{V}}(\langle s_{j-1}, h_{j-1} \rangle; \langle s_j, h_j \rangle)$, $I_V(\langle s_{j-1}, h_{j-1} \rangle; \langle s_j, h_j \rangle)$, $j = 2, 3, \ldots, n$, are unknown.

N1: Fix $\langle t, k \rangle$, $t \in M$. If $\overline{\tau}_t(k) = \infty$, then A3 implies that $\tau_t(k) = \infty$, i.e., t never fires for the kth time in either the unperturbed or perturbed net. Hence, we can set $\Delta_t(k) = \infty$, which is known. Suppose that $\overline{\tau}_t(k) < \infty$. Either $\langle t, k \rangle$ is an initial firing, or it is not. When $\langle t, k \rangle$ is an initial firing,

$$\Delta_t(k) = \begin{cases} \tau_t(1) - \overline{\tau}_t(1) & k = 1 \\ \left(\tau_t(k) - \overline{\tau}_t(k) \right) - \left(\tau_t(k-1) - \overline{\tau}_t(k-1) \right) & k > 1 \end{cases} \tag{7}$$

is known by the necessity hypothesis and the definition of M. When $\langle t, k \rangle$ is not an initial firing, by Lemma 1 there exists a rooted token track $S = \{\langle s_i, h_i \rangle\}_{i=1}^n$, with $n > 1$, such that $\langle s_n, h_n \rangle = \langle t, k \rangle$. It suffices to demonstrate the existence of a V and $\overline{V}$ such that $\Delta_{s_n}(h_n)$ is known. Clearly, this follows from $A3$ - $A5$, when $\overline{I}_{\overline{V}}(\langle s_{b-1}, h_{b-1} \rangle; \langle s_b, h_b \rangle) = I_V(\langle s_{b-1}, h_{b-1} \rangle; \langle s_b, h_b \rangle) = 0$, $b = 2, 3, \ldots, n$.

N2: Fix the rooted token track $S = \{\langle s_i, h_i \rangle\}_{i=1}^n \in D_{\mathcal{P}}$. It suffices to demonstrate a $\overline{V}$ and V such that $\overline{I}_{\overline{V}}(\langle s_{b-1}, h_{b-1} \rangle; \langle s_b, h_b \rangle)$ and $I_V(\langle s_{b-1}, h_{b-1} \rangle; \langle s_b, h_b \rangle)$ are known for all $b = 2, 3, \ldots, n$. For all $b \in \{2, 3, \ldots, n\}$, $A3$ - $A5$ imply that there exists a V and a $\overline{V}$ such that S exists, $\overline{I}_{\overline{V}}(\langle s_{i-1}, h_{i-1} \rangle; \langle s_i, h_i \rangle) = 0$, for $i = 2, 3, \ldots, b - 1, b + 1, \ldots, n$ and $I_V(\langle s_{i-1}, h_{i-1} \rangle; \langle s_i, h_i \rangle) = 0$, for $i = 2, 3, \ldots, n$. Because $\Delta_{s_n}(h_n)$ is known by $N1$, $\overline{I}_{\overline{V}}(\langle s_{b-1}, h_{b-1} \rangle; \langle s_b, h_b \rangle)$ is known. Similarly, interchanging $I_V(\cdot)$ and $\overline{I}_{\overline{V}}(\cdot)$ we can conclude that $I_V(\langle s_{b-1}, h_{b-1} \rangle; \langle s_b, h_b \rangle)$ is known, for all $b = 2, 3, \ldots, n$.

Sufficiency: Fix $\overline{V}$ and V. Assume that $\Delta_t(k)$ is known for all $t \in M$ and $k \in \mathbb{N}$ and that the idle times $\overline{I}_{\overline{V}}(\langle s_{b-1}, h_{b-1} \rangle; \langle s_b, h_b \rangle)$, $I_V(\langle s_{b-1}, h_{b-1} \rangle; \langle s_b, h_b \rangle)$, $b = 2, 3, \ldots, n$, are known for $S = \{\langle s_i, h_i \rangle\}_{i=1}^n \in D_{\overline{V}}$. It suffices to show that $\tau_t(k)$ is known for all $t \in M$, $k \in \mathbb{N}$.

Fix $t \in M$, $k \in \mathbb{N}$. For all transition firings $\langle t, k \rangle$ with $\overline{\tau}_t(k) = \infty$, $A3$ implies that $\tau_t(k) = \infty$ is known. Assume that $\overline{\tau}_t(k) < \infty$. Because the net is FIFO, when $\langle t, k \rangle$ is an initial firing,

$$\tau_t(k) = \overline{\tau}_t(k) + \sum_{i=1}^k \Delta_t(i) \tag{8}$$

is known by the sufficiency hypothesis and the definition of M. Assume that $\langle t, k \rangle$ is not an initial firing. By Lemma 1, there exists a rooted token track $S = \{\langle s_i, h_i \rangle\}_{i=1}^n$, with $n > 1$, such that $\langle t, k \rangle = \langle s_n, h_n \rangle$.

Next we must show that we can use this rooted token track to find $\tau_t(k)$. Consider (4) written for S. Either $\langle s_1, h_1 \rangle$ is an initial firing, or it is not. When $\langle s_1, h_1 \rangle$ is an initial firing, by a rearrangement of (8) (with $\langle t, k \rangle$ replaced by $\langle s_1, h_1 \rangle$), $A2$, the sufficiency hypothesis, and the definition of M, $\tau_t(k) = \tau_{s_n}(h_n)$ is known.

Suppose that $\langle s_1, h_1 \rangle$ is not an initial firing. If we can show that $\overline{\tau}_{s_1}(h_1) - \tau_{s_1}(h_1)$ is known, it follows from $A2$, the sufficiency hypothesis, and the definition of M, that $\tau_t(k) = \tau_{s_n}(h_n)$ is known. Because $\langle s_1, h_1 \rangle$ is not an initial firing, $s_1 \in M$. By Lemma 1 there exists a rooted token track S' terminating at $\langle s_1, h_1 \rangle$. Because $s_1 \in M$, we can apply the same reasoning to S' that we did to S. Because all tokens trace their origins to some initial firing, if we do this recursively, we must eventually find a S' for which the first firing is an initial firing. Hence, we can find $\tau_t(k)$. ∎

Theorem 1 characterizes the timing information required to solve TP0. In essence, it states that we have to know the perturbations incurred by the monitors, and that we have to be able to track perturbations when they interact. To track perturbations we must, in turn, know the idle times of tokens in the net.

Theorem 1 does not specify how we should acquire this information. When this information can not be derived from net structure we are forced to derive it from the monitor firing times. This is discouraging when tokens can idle in any place (e.g., in TPN models of queueing networks) because it implies that every transition firing time must be derivable from monitor firing times. This, in turn, implies that perturbation tracking in these systems is roughly equivalent to resimulating the system, i.e., every place in the net contributes a term to (4).

Fortunately, in many important systems tokens can not idle in every place. Net structure imposes constraints on the locations of non-zero idle times (e.g., some idle times are known to be zero for all possible firing delay sequences V), and consequently, it is not necessary to know the firing times of many transitions for perturbation tracking, i.e., $M \subset T$. For instance, in the TPN models of the distributed message-passing system software that motivated this work, non-zero idle times can only occur at places immediately upstream from the transitions that synchronize the message passing. Accordingly, it is sufficient to know the firing times of transitions in the vicinity of these synchronizing transitions.

6. SUFFICIENT MONITOR CONDITIONS

In this section, we use assumptions $A1$ - $A5$ to identify conditions on the placement of monitors sufficient to satisfy Theorem 1. An example of a trivial sufficient condition is to set $M = T$. Given $N1$, this ensures that both the arrival time of each token and each transition firing delay are known. Additional information derived from structure, such as the knowledge that a subnet of the TPN is *safe*, may allow tracking perturbations with fewer monitors. Here, we present sufficient monitor conditions for $N1$ and $N2$ that exploit the safeness of places to ensure that we can track perturbations without knowing all transition firing times. Specifically, we restrict attention to TPNs with the following property:

P: For all rooted token tracks $\{\langle s_i, h_i \rangle\}_{i=1}^{n} \in D_{\mathcal{P}}$, every $p \in \mathcal{I}(s_i)$, $i = 2, 3, ..., n$, is safe, i.e., places along rooted token tracks contain at most one token.

Note that P is trivially satisfied when the net is safe.

Theorem 2 *For a TPN with fixed $\overline{V}$ and V, satisfying assumptions* A1 - A5 *and* P, N1 *is satisfied if*

(1) *for all* $t \in M$, *there exists a* $u \in M$ *such that* $\Delta_t(k) = \Delta_u(\ell)$, *for some* $k, \ell \in \mathbb{N}$, *and either* $\langle t, k \rangle \in C_{\overline{V}}(\langle u, \ell \rangle)$ *or* $\langle u, \ell \rangle \in C_{\overline{V}}(\langle t, k \rangle)$, *i.e., monitors occur in pairs and have equal perturbations;*

(2) *for all* $t, u \in M$, *and* $k, \ell \in \mathbb{N}$ *such that* $\langle t, k \rangle \in C_{\overline{V}}(\langle u, \ell \rangle)$, $I_V(\langle t, k \rangle; \langle u, \ell \rangle) = \overline{I}_{\overline{V}}(\langle t, k \rangle; \langle u, \ell \rangle) = 0$; *and*

(3) *for all* $t \in M$, $v_t = 0$.

Proof: Fix $t, u \in M$ and $k, \ell \in \mathbb{N}$ such that conditions (1) - (3) of the theorem are satisfied. Without loss of generality, suppose that $\langle t, k \rangle \in C_{\overline{V}}(\langle u, \ell \rangle)$. By definition, there exists a rooted token track $S = \{\langle t, k \rangle, \langle u, \ell \rangle\} \in D_{\overline{V}}$. Writing (4) for S we get, after rearrangement

$$\overline{\tau}_u(\ell) - \tau_u(\ell) - \left(\overline{\tau}_t(k) - \tau_t(k) \right) = -\Delta_u(\ell). \tag{9}$$

Because $v_t = 0$, $\tau_u(\ell) - \tau_t(k) = v_t = 0$, and (9) becomes

$$\overline{\tau}_u(\ell) - \overline{\tau}_t(k) = -\Delta_u(\ell), \tag{10}$$

which is known by the definition of M. Hence $\Delta_t(k) = \Delta_u(\ell)$ is known. Because every transition $t \in M$ satisfies conditions (1) - (3), $\Delta_t(k)$ is known for all $t \in M$ and $k \in \mathbb{N}$.   ∎

If monitors occur in pairs, have no intervening idle times, and have identical induced perturbations equal to their firing delay (i.e., the monitor firing delay is perturbed to zero), by Theorem 2, the perturbation can be found from the monitor transition firing times. In particular, these requirements can be satisfied for TPNs modeling software for distributed message-passing systems [5].

Theorem 3 N2 *is satisfied for a TPN with fixed* $\overline{V}$ *and* V, *satisfying* N1, *assumptions* A1 - A5, *and property* P, *when* $M = \{q : q \in \mathcal{I}(\mathcal{I}(t)), \mid \mathcal{I}(t) \mid > 1\}$.

Proof: Fix $\overline{V}, V$, and $\{\langle s_i, h_i \rangle\}_{i=1}^n \in D_{\overline{V}}$ for some $n > 1$. It suffices to show that the idle times $\overline{I}_{\overline{V}}(\langle s_{j-1}, h_{j-1} \rangle; \langle s_j, h_j \rangle)$ and $I_V(\langle s_{j-1}, h_{j-1} \rangle; \langle s_j, h_j \rangle)$ are known for all $j \in \{2, 3, \ldots, n\}$. Fix $j \in \{2, 3, \ldots, n\}$. There are two cases to consider: $\mid \mathcal{I}(s_j) \mid = 1$ and $\mid \mathcal{I}(s_j) \mid > 1$.

When $\mid \mathcal{I}(s_j) \mid = 1$, safeness ensures that no token arrives during a transition firing and we can write

$$\overline{I}_{\overline{V}}(\langle s_{j-1}, h_{j-1} \rangle; \langle s_j, h_j \rangle) = I_V(\langle s_{j-1}, h_{j-1} \rangle; \langle s_j, h_j \rangle) = 0. \tag{11}$$

When $\mid \mathcal{I}(s_j) \mid > 1$, $\mathcal{I}(\mathcal{I}(s_j)) \subseteq M$; hence, by definition 2 $j = 2$, and we can write

$$\overline{I}_{\overline{V}}(\langle s_1, h_1 \rangle; \langle s_2, h_2 \rangle) = \max_{\langle t, k \rangle \in C_{\overline{V}}(\langle s_2, h_2 \rangle)} \left\{ \overline{\tau}_t(k) - \overline{\tau}_{s_1}(h_1) \right\} \tag{12}$$

and

$$I_V(\langle s_1, h_1 \rangle; \langle s_2, h_2 \rangle) = \max_{\langle t, k \rangle \in C_V(\langle s_2, h_2 \rangle)} \left\{ \tau_t(k) - \tau_{s_1}(h_1) \right\}. \tag{13}$$

Because $\langle s_1, h_1 \rangle \in C_{\overline{V}}(\langle s_2, h_2 \rangle)$, and $C_{\overline{V}}(\langle s_2, h_2 \rangle) = C_V(\langle s_2, h_2 \rangle)$ (by $A3$), it suffices to show that $\overline{\tau}_t(k)$ and $\tau_t(k)$ are known for all $\langle t, k \rangle \in C_{\overline{V}}(\langle s_2, h_2 \rangle)$. Fix $\langle t, k \rangle \in C_{\overline{V}}(\langle s_2, h_2 \rangle)$. Note that $t \in M$, so $\overline{\tau}_t(k)$ is known by definition. There are two possibilities. Either $\langle t, k \rangle$ is an initial firing, or it is not. If $\langle t, k \rangle$ is an initial firing, we can find $\overline{\tau}_t(k)$ using (8).

Suppose that $\langle t, k \rangle$ is not an initial firing. Because $t \in M$, Lemma 1 implies that there exists a rooted token track S' terminating at $\langle t, k \rangle$. We can apply the same reasoning to S' that we did to S. Because all tokens trace their origin to some initial firing, if we do this recursively, we must eventually find a S' for which the first firing $\langle t, k \rangle$ is an initial firing and, by a rearrangement of (8), $\overline{\tau}_t(k) - \tau_t(k)$ is known. Accordingly, by $A2$, $\tau_t(k)$ can be found for all $\langle t, k \rangle \in C_{\overline{V}}(\langle s_2, h_2 \rangle)$. It follows that $\overline{I}_{\overline{V}}(\langle s_1, h_1 \rangle; \langle s_2, h_2 \rangle)$ and $I_V(\langle s_1, h_1 \rangle; \langle s_2, h_2 \rangle)$ are known.   ∎

Theorems 2 and 3 describe one monitor placement that can provide the timing information required by Theorem 1. In this placement, all $t \in T$ for which

$\{\langle t, k \rangle\} = C_V(\langle u, \ell \rangle)$, for some $u \in T$ and all k, $\ell \in \mathbb{N}$, are not monitor transitions; hence, perturbations can be tracked without knowing all transition firing times.

Of course, this monitor placement is only sufficient. Theorem 1's conditions can also be satisfied by structural information, or by other monitor sets M. For instance, perturbations Δ_t, $t \in M$, may be known for some TPNs. In the software monitoring problem that motivates our study, the perturbations Δ_t, $t \in M$, are known when the software probes report the duration of their execution.

Although Theorems 2 and 3 provide conditions sufficient to ensure that perturbation tracking is possible, they do not provide an algorithm. One approach would be to proceed as follows.

Assume, for fixed $\overline{V}$, that the firing times $\overline{\tau}_t$ are given for all $t \in M$ and that the conditions of Theorem 1 are satisfied.

Step 1: Construct a set $K \subseteq D_{\overline{V}}$ consisting of all rooted token tracks $\{\langle s_i, h_i \rangle\}_{i=1}^n$, $n > 1$, for which $\langle s_1, h_1 \rangle$ is an initial firing. Let K_j denote the jth of these tracks. Because the conditions of Theorem 1 are satisfied, we can use (8) to find $\overline{\tau}_{s_1}(h_1) - \tau_{s_1}(h_1)$, and (4) to find $\overline{\tau}_{s_n}(h_n)$, and $\tau_{s_n}(h_n)$, where n is the length of K_j. Hence, for all $K_j \in K$, the cumulative perturbations $\overline{\tau}_{s_1}(h_1) - \tau_{s_1}(h_1)$ and $\overline{\tau}_{s_n}(h_n) - \tau_{s_n}(h_n)$ are known.

Step 2: Choose any rooted token track $R = \{\langle r_i, g_i \rangle\}_{j=1}^m \notin K$, with $m > 1$. By the construction of K, track R must begin with a monitor firing $\langle r_1, g_1 \rangle$. If $\langle r_1, g_1 \rangle$ belongs to some rooted token track in K, then, by definition 2, it must begin or terminate that track. It follows, due to step 1, that $\overline{\tau}_{r_1}(g_1) - \tau_{r_1}(g_1)$ is known. Hence, we can use (4) to find $\overline{\tau}_{r_m}(g_m) - \tau_{r_m}(g_m)$. Moreover, we can add R to K without destroying the property that, for all $K_j = \{\langle s_i, h_i \rangle\}_{i=1}^n \in K$, the cumulative perturbations $\overline{\tau}_{s_1}(h_1) - \tau_{s_1}(h_1)$ and $\overline{\tau}_{s_n}(h_n) - \tau_{s_n}(h_n)$ are known. When $\langle r_1, g_1 \rangle$ does not belong to any token track in K, select another rooted token track $R \notin K$. By repeating this procedure we eventually add all rooted token tracks to K, so that, by Lemma 1 and A2, $\tau_t(k)$ will be known for all $\langle t, k \rangle$, with $t \in M$ and $k \in \mathbb{N}$ such that $\overline{\tau}_t(k) < \infty$.

For a more detailed description of this algorithm, see [5].

7. CONCLUSION

In this paper we have characterized the timing information required to track timing perturbations in TPNs in which firing times are only known for a set of monitor transitions that are assumed to produce the timing perturbations. Specifically, we have shown that for perturbation tracking it is necessary and sufficient to know the duration of the induced perturbations, and the idle times tokens incur at places.

In the worst case, this information must be derived solely from monitor transition firing times. Often, some information can also be derived from TPN structure. For instance, in safe nets non-zero idle times can not occur in arbitrary locations; hence fewer monitors are required to track perturbations. In this case, we have shown that it is sufficient to know the induced perturbations and to require that all transitions immediately upstream from multi-input transitions be monitors.

Presently, our results are being used to develop practical algorithms for tracking the timing perturbations introduced when software probes are used to monitor the internal state of executing, asynchronous, distributed software. The relationship between these algorithms and perturbation analysis will be considered in a subsequent paper.

REFERENCES

[1] Y.-C. Ho, X.-R. Cao, and C. Cassandras, "Infinitesimal and finite perturbation analysis for queueing networks," *Automatica*, Vol. 19, No. 4, April 1983, pp. 439-445.

[2] Y.-C. Ho and X.-R. Cao, *Perturbation Analysis of Discrete Event Dynamic Systems*, Kluwer Academic Publishers, 1991.

[3] M. S. Andersland and T. L. Casavant, "Recovering uncorrupted event traces from corrupted event traces in parallel/distributed computing systems," *Proc. of the 1991 International Conference on Parallel Processing*, August 1991, pp. II.108-II.116.

[4] J. E. Lumpp, Jr., J. A. Gannon, M. S. Andersland, and T. L. Casavant, "A technique for recovering from software instrumentation intrusion in message-passing systems," *University of Iowa Electrical and Computer Engineering Technical Report* TR-ECE-920817, August 1992.

[5] J. A. Gannon, J. E. Lumpp, Jr., K. J. Williams, M. S. Andersland, and T. L. Casavant, "Tracking timing perturbations in timed Petri nets," *University of Iowa Electrical and Computer Engineering Technical Report* TR-ECE-920304, March 1992.

[6] C. G. Cassandras, *Discrete Event Systems: Modeling and Performance Analysis*, Irwin, Inc., and Aksen Associates Inc., 1993.

[7] F. Baccelli, G. Cohen, and B. Gaujal, "Evolution equations of timed Petri nets," *Proc. of the 30th IEEE Conference on Decision and Control*, Vol. 2, December 1991, pp. 1139-1144.

[8] H. P. Hillion and J.-M. Proth, "Performance evaluation of job-shop systems using timed event-graphs," *IEEE Transactions on Automatic Control*, Vol. 34, No. 1, January 1989, pp. 3-9.

EXTENSIONS TO THE THEORY OF OPTIMAL CONTROL OF DISCRETE EVENT SYSTEMS*

Raja Sengupta   Stéphane Lafortune[†]

1. INTRODUCTION AND PROBLEM FORMULATION

This paper describes new results concerning the optimal control problem (OCP) formulated in [6]. The discrete event system (DES) is modeled as a finite vertex directed graph, denoted by $G = < V_G, E_G, v_0, v_m >$, where V_G is the finite set of vertices of G, E_G the set of directed edges, v_0 the unique initial vertex, and v_m the unique terminal vertex; it is also assumed that G has at most one edge between a pair of vertices and that all vertices in G are accessible with respect to v_0 and co-accessible with respect to v_m. We call a graph G with such properties an *admissible graph*. In general we will use the functional notation $E_f(.)$ and $V_f(.)$ to denote the set of edges and the set of vertices respectively of an admissible graph. Any path on this graph represents a possible behavior of the DES. Intuitively, paths starting at v_0 and ending at v_m may be viewed as complete or properly terminating behaviors. Accordingly it is assumed that no non-terminating paths are admissible in the controlled behavior. The admissible paths correspond to the language marked by G, denoted by $L_m(G)$. In [6] the same problem was treated but with many restrictions on G. Principal amongst these was the requirement that G be acyclic. This critical qualification is now dispensed with albeit with some increase in the computational complexity.

It has been assumed in [5] that events are generated by a DES in a manner that is spontaneous, asynchronous and instantaneous. We adapt the same to our graph-theoretic model. Thus the DES will spontaneously execute any one possible path in a given time line. Similarly control will be used to restrict the set of possible behaviors of the system and will not, in general, force one particular behavior. Control action represents the removal of edges from the graph representing the uncontrolled DES. In other words control is a map defined at each vertex of the uncontrolled DES which specifies the edges to be removed. Accordingly a controlled DES is a subgraph of the uncontrolled DES and the space of controlled DES is bijectively identifiable with the set of subgraphs of the uncontrolled system. Our optimization problem is thus formulated over the set of subgraphs of a given uncontrolled DES.

Past work in the area of cost oriented optimal control for DES consists of [3, 1, 2, 4]. Our formulation and results either subsume or are substantially

*Research supported in part by the National Science Foundation under grant ECS-9057967, with additional support from DEC and GE.

†The authors are with the Department of Electrical Engineering and Computer Science, University of Michigan, Ann Arbor, MI 48109-2122, USA; e-mail: {raja, stephane}@eecs.umich.edu

different from these three approaches. Our OCP draws its rationale from the intuition of classical optimal control theory. Two positive costs are defined on the set of edges of the DES. They are the path and control costs respectively. The cost associated with a terminating behavior or path of length n, $(p = e_1 \ldots e_n)$ lying in a controlled system A, where A is a subgraph of G, is denoted by $c(p, A)$ and defined to be

$$c(p, A) \;=\; \sum_{i=1}^{i=n} c_p(e_i) + \sum_{i=0}^{i=n} \sum_{(v_i,v) \in E_G - E_f(A)} c_c((v_i, v))$$

where $e_i \;=\; (v_{i-1}, v_i)$.

In other words the cost of a path p lying in a subgraph A is (i) the sum of the path cost $c_p(e)$ of each edge e in the path together with, (ii) the sum of the control cost $c_c(e)$ of each edge e that lies in the uncontrolled DES G but not in the controlled DES A, and is attached to a vertex visited by the path p. Thus if a path lies in two graphs, one of which is a subgraph of the other, then the cost of the path is lower in the bigger subgraph. This represents the fundamental tradeoff in optimal control theory between costs on system behavior and costs on control. Any attempt to lower worst case path costs by control action raises the control costs. Thus the optimal control will generally not force the shortest path, but instead define some subgraph consisting of several paths.

The idea of a positive control cost is motivated by our interpretation of control. The control objective of the supervisor is to drive the system to the terminal state v_m which represents some completed task or terminated process. To fulfill this objective in minimum time or space (depending on the interpretation of path cost), the supervisor disables certain events. We view this as impacting on the environment in which the system operates. The environment is inhibited by the action of the supervisor and consequently cost is incurred. For instance the disabling of events may obstruct other processes occurring in the environment of the system for which the supervisor is not responsible. The cost of control would then represent a penalty for such interference. For example control actions which isolate parts of a computer network might incur control cost penalties arising out of user inconvenience.

It is apparent that the exact cost incurred by a controlled system is indeterminate in the sense that it depends on its dynamic behavior which is spontaneous. However each subgraph A of G has a unique supremal or worst case cost

$$c_{\text{sup}}(A) = \sup_{p_m \in L_m(A)} c(p_m, A)$$

that represents the worst case case which may be incurred by the system A over all its possible terminating paths. We aim to minimize this worst case over the set of all constructible subgraphs of the uncontrolled DES. More precisely, the problem formulation is as follows.

Let us call $A =< V_f(A), E_f(A), v_a, v_b >$ an *admissible subgraph* of G rooted at v_d if (i) A is itself an admissible graph (as defined at the beginning of this section), (ii) A is a subgraph of G, and (iii) $v_a = v_d$ and $v_b = v_m$. Consider the set of all admissible subgraphs of G rooted at v_0. Let this set be denoted by $\mathcal{S}(G, v_0)$. We wish to find $A_0 \in \mathcal{S}(G, v_0)$ such that

$$c_{\text{sup}}(A_0) = \min_{A \in \mathcal{S}(G, v_0)} c_{\text{sup}}(A).$$

It may be noted that we search over the entire set $\mathcal{S}(G, v_0)$. The definition of admissibility ensures that this set is finite. Moreover it is assumed that each member of $\mathcal{S}(G, v_0)$ is constructible or there exists some control law which, will yield the particular subgraph. Thus, any edge can be removed or equivalently there is total controllability.

We conclude this section with an observation on the effect of uncontrollability. Since all costs are positive the optimal subgraph must be acyclic. The presence of uncontrollable edges may make all acyclic subgraphs unconstructible. But so long as even one acyclic graph is constructible a solution to the OCP exists. The optimal solution is generally not unique. But just as in the Ramadge & Wonham paradigm there exists a minimally restrictive optimal solution.

2. PRINCIPAL RESULTS

We define A_0 (an admissible subgraph of G) to be DP-optimal if

(i) A_0 is optimal, and

(ii) $\forall v \in V_f(A_0)$, $c_{\text{sup}}(S(A_0, v)) = \min_{B \in \mathcal{S}(G, v)} c_{\text{sup}}(B)$

where $S(A, v)$ represents the maximal admissible subgraph of A rooted at v. $S(A, v)$ is maximal in the sense that every other admissible subgraph of A rooted at v is an admissible subgraph of $S(A, v)$. We assume that the symbol $\mathcal{S}_D^o(G, v)$ will denote the set of DP-optimal subgraphs of G rooted at v.

The content of (ii) above is that every maximal subgraph of a DP-optimal solution is itself DP-optimal in its appropriate class or $\mathcal{S}(G, v)$ for some $v \in V_f(A_0)$. The results in [6] generalize to the problem at hand to show that this class is not vacuous. All optimal solutions do not have optimal substructure, hence the need to consider the class of DP-optimal solutions. Moreover it is only within this class of DP-optimal solutions that it is possible to establish the existence of a maximal (or minimally restrictive) solution. This is established by the following theorem which shows that the merger (denoted by the symbol $\oplus$) of DP-optimal solutions is DP-optimal.

Theorem 1 *Let* $S_a^o \in \mathcal{S}_D^o(G, v_d)$ *and* $S_b^o \in \mathcal{S}_D^o(G, v_d)$. *Then* $S_a^o \oplus S_b^o \in \mathcal{S}_D^o(G, v_d)$.
□

Since the set $\mathcal{S}(G, v_d)$ is finite we get the existence of a unique maximal element by induction.

The maximal DP-optimal solution can be computed by a backward recursive dynamic programming algorithm. The algorithm starts at v_m and will terminate after processing the vertex v_0. At each step it is necessary to perform two minimizations. One is linear in $\|V_G\|$ and the other at first glance appears to be exponential. The following theorem shows how the minimization can be reduced to the complexity of a merge-sort.

Before stating the theorem it is necessary to extend the notation. A one-step subgraph of G rooted at v_d is defined as any subgraph whose edges are contained in the set $\{(v_d, v) : (v_d, v) \in E_G\}$. $\mathcal{S}_1(G, v_d)$ denotes the set of all one-step subgraphs of G rooted at v_d. A dynamic programming equation, stating the problem as a minimization on one-step subgraphs, may be formulated as follows.

$$c_{\text{sup}}(M_D^o(G, v_d)) = \min_{S' \in \mathcal{S}_1(G, v_d)} [\max_{(v_d, v) \in E_f(S')} [c((v_d, v), S') + c_{\text{sup}}(M_D^o(G, v))]]$$

$$\text{where} \quad c((v_d, v), S') = c_p((v_d, v)) + \sum_{(v_d, v') \in E_G - E_f(S')} c_c((v_d, v'))$$

$$M_D^o(G, v) = \text{Maximal DP-optimal solution of } G \text{ rooted at } v.$$

The number of possible one-step subgraphs is exponential in the number of vertices of G. However the minimizing S' may be found as follows. Let it be known that the edges of the minimizing S' lie possibly in a set $E_0 = \{e_1, \ldots, e_n\}$. Such a set E_0 occurs in step (iv) of subsection 3.3 of the Algorithm. Order the set E_0 so that

$$i < j \Rightarrow c_p(e_i) + c_{\text{sup}}(M_D^o(G, v_i)) \leq c_p(e_j) + c_{\text{sup}}(M_D^o(G, v_j))$$
$$\text{where } e_k = (v_d, v_k).$$

Consider the sequence of graphs

$$A_k = A_k' \oplus (\oplus_{i=1}^{i=k} M_D^o(G, v_i))$$
$$\text{where} \quad A_k' = \text{the graph defined by } \{e_1, \ldots, e_k\} \text{ and}$$
$$c_{\text{sup}}(A_k) = c_p(e_k) + c_{\text{sup}}(M_D^o(G, v_k)) + \sum_{i=k+1}^{i=n} c_c(e_i) + \sum_{e \in E_G - E_0} c_c(e).$$

The theorem may now be stated as

Theorem 2 *Let* $< A_k >_{k=1}^{k=n}$ *be as above. If* A_r *is the greatest* r *such that*

$$c_{\text{sup}}(A_r) = \min_{1 \leq k \leq n} c_{\text{sup}}(A_k)$$

then $M_D^o(G, v_d) = A_r$. □

Observe that the complexity of the sort is $O(n \log n)$. All other search operations are linear. The next theorem summarizes our results.

Theorem 3 *For all* $v \in V_f(G)$:

(i) *the unique maximal DP-optimal subgraph of* G *rooted at* v *exists.*

(ii) $M_D^o(G, v)$ *is computable in* $O(n^3 \log n)$, *where* $\|V\| = n$. □

We discuss the solution algorithm below. The proofs of its correctness and complexity suffice to prove the above theorem. They are omitted in this paper.

3. ALGORITHM

We present in this section an algorithm to compute the maximal optimal and admissible solution of an admissible graph G. All new notation adopted in this section is defined below for easy reference.

(i) SL = Solved list of vertices (also sometimes called the closed list).

(ii) $E_{opt}(v)$ = The set of edges of $S_1(M_D^o(G,v),v)$, the maximal one-step subgraph of $M_D^o(G,v)$ rooted at v.

(iii) $CL = \{(v, c_{\sup}(M_D^o(G,v))) : v \in V_G\}$. This is the cost list maintained by the algorithm for the recursive computation of the cost associated with a particular subgraph.

Note that the variables CL^{temp} and $E_{opt}^{temp}(.)$ are also used for temporary storage of the same quantities.

(iv) C = Set of vertices to be processed in the current iteration.

(v) $P_f(C) = \{v \in V_G : (v, v') \in E_G, v' \in C\}$.

(vi) $S_f(v) = \{v' \in V_G : (v, v') \in E_G\}$. This is the set of successor vertices of the vertex v.

For the convenience of modularity we present the algorithm as a main program and two subprograms referred to as Optimize and One-step optimize respectively. The main program primarily orders the backward recursive search and updates C and SL based on data provided by the sub-program Optimize. The optimization at each vertex is described in the subprogram Optimize which in turn calls the subprogram One-step optimize.

3.1   Explanation of the algorithm

This explanation of the solution algorithm is patterned on Dijkstra's shortest path algorithm since both are graph based backward recursive dynamic programming algorithms.

The Dijkstra algorithm will start at the terminal vertex and place it on the solved list (SL). At each step it will develop the set of parents of SL (denoted by C). From this set the algorithm identifies the next vertex to be added to SL. This identification is $O(n)$ and since the identification process is repeated each time a vertex is added to SL the overall complexity is $O(n^2)$.

This algorithm also starts at the terminal vertex (v_m) and places it in SL. At each step it develops C (3.2, vii) and identifies the next vertex to be added to SL (3.2, iv & v). Unlike the Dijkstra algorithm the complexity of the identification process is $O(n^2 \log n)$. Since the process is repeated each time a vertex is added to SL the overall complexity is $O(n^3 \log n)$.

We now explain the additional steps which result in the complexity $O(n^2 \log n)$. The next vertex to be added to SL is determined as follows. Develop the set C and pick some $v \in C$. In general v has edges leading to children which are not in SL. Disable all such edges and consider the one-step subgraph constituted of the remaining edges (3.3, iv). This is the set E_0. Sort these edges as explained in section 2 and construct the sequence of subgraphs $< A_j >$ referred to in Theorem 2. This sort is $O(n \log n)$. Next find the minimum cost A_j (refer to 3.4). Let it be denoted by A_v. The complexity thus far is

$$O(n \log n + n) = O(n \log n).$$

Since A_v must be found for each $v \in C$ the overall complexity is $O(n^2 \log n)$. The vertices realizing $\min_{v \in C} c_{\sup}(A_v)$ may be added to SL with the corresponding A_v representing $M_D^o(G, v)$ (3.2, v). Finding this minimum is $O(n)$ which keeps the overall complexity at

$$O(n^2 \log n + n) = O(n^2 \log n).$$

This completes the description of the algorithm for the computation of the maximal DP- optimal solution for cyclic graphs.

3.2 The main program

(i) Input: V_G, E_G, v_o, v_m.

(ii) Initialize: $C = \{v_m\}, SL = \emptyset, CL = \emptyset, CL^{temp} = \emptyset$.

(iii) Optimize: Call subprogram Optimize with argument C.

(iv) Compute

$$A = \{v_d \in C : c_{\max}(E_{opt}^{temp}(v_d)) = \min_{v \in C} c_{\max}(E_{opt}^{temp}(v))\}.$$

(Note: $c_{\max}$ is computed in the subprogram One-step optimize.)

(v)

$$
\begin{aligned}
\forall v \in A \quad E_{opt}(v) &= E_{opt}^{temp}(v) \\
CL &\rightarrow CL \cup \{(v, c_{\max}(E_{opt}(v))\} \\
SL &\rightarrow SL \cup A \\
CL^{temp} &= \emptyset.
\end{aligned}
$$

(vi) Termination Condition: Is $v_o \in SL$? If yes then STOP. Otherwise continue.

(vii) Computation of the vertices to be optimized in the next iteration:

$$C = P_f(SL) - SL.$$

(viii) GOTO (iii).

3.3 Optimize

(i) Input: C

(ii) Pick any $v_d \in C$.

(iii) Update C: $C = C - \{v_d\}$.

(iv) Compute

$$E_0 = E_f(S_1(G, v_d)) - \{(v_d, v) : v' \notin SL\} \;=\; \{e_1, \ldots, e_n\},$$

where $e_i = (v_d, v_i)$ and order the e_i such that

$$i < j \Rightarrow c_p(e_i) + c_{\text{sup}}(M_D^o(G, v_i)) \;\leq\; c_p(e_j) + c_{\text{sup}}(M_D^o(G, v_j)).$$

(v) If $E_o = \emptyset$ then set $c_{\max}(E_o) = \sum_{(v_d,v) \notin E_o} c_c(v_d, v)$. Else set

$$c_{\max}(E_o) = c_p(e_n) + c_{\text{sup}}(M_D^o(G, v_n)) + \sum_{(v_d,v) \notin E_o} c_c(v_d, v).$$

(vi) Call subprogram One-step optimize.

(vii) Termination condition: Is $C = \emptyset$? If yes then return to the main program. Otherwise GOTO (iii).

3.4   One-step optimize

In this section we use the notation $c_{\max}(E)$ where $E = \{e_1, \ldots, e_j\}$ to denote the following calculation:

$$c_{\max}(E) = c_p(e_j) + c_{\text{sup}}(M_D^o(G, v_j)) + \sum_{i=j+1}^{i=n} c_c(e_i) + \sum_{(v_d,v) \notin E_o} c_c(v_d, v).$$

where it is assumed that $E_0 = \{e_1, \ldots, e_n\}$ is ordered by increasing cost as described in section 3.3. The aim is to find the A_j that will satisfy Theorem 2.

(i) Input: E_0, $c_{\max}(E_0)$

(ii) Initialize: $E = E_0$, $E' = E_0$, $CMAX = c_{\max}(E_0)$, $E_{opt}^{temp}(v_d) = \emptyset$.

(iii) Compute $E' \leftarrow E' - \{e_i : i = \max_{e_j \in E'} j\}$. If $E' \neq \emptyset$ then set

$$c_{\max}(E') = c_p(e_{i-1}) + c_{\text{sup}}(M_D^o(G, v_{i-1})) + \sum_{k=i}^{k=n} c_c(e_k) + \sum_{(v_d,v) \notin E_o} c_c(v_d, v).$$

(iv) Termination condition: Is $E' = \emptyset$?

    If yes then set

$$E_{opt}^{temp}(v_d) \;\leftarrow\; E_{opt}^{temp}(v_d) \cup E,$$
$$CL^{temp} \;\leftarrow\; CL^{temp} \cup \{(v_d, c_{\max}(E))\}$$

and return to Optimize. Otherwise continue.

(v) Recursion condition: Is $c_{\max}(E') < CMAX$? If not then GOTO (iii). If yes then continue.

(vi) Set $E = E', CMAX = c_{\max}(E')$.

(vii) GOTO (iii).

4. CONCLUSION

The optimal control problem discussed in this paper and in [6] appears to be a general framework within which most of existing theory can be absorbed and several new and interesting problems formulated. Here we have been able to relax the restrictive assumption of [6] that G be acyclic. The price paid in terms of computational complexity for this generalization is reasonable: from $O(n^2 \log n)$ to $O(n^3 \log n)$. The only restrictive assumption that remains on the structure of G is the condition that there be a unique terminal vertex. At present, we are unable to guarantee solutions in polynomial time (in terms of the number of vertices of G) if it be relaxed.

REFERENCES

[1] Y. Brave and M. Heymann. On optimal attraction of discrete-event processes. Center for Intelligent Systems Report 9010, Technion, Haifa, Israel, October 1990.

[2] Y. Brave and M. Heymann. Stabilization of discrete event processes. *Int. J. Control*, 51(5):1101–1117, 1990.

[3] R. Kumar and V. Garg. Optimal control of discrete event dynamical systems using network flow techniques. Preprint, Department of Electrical and Computer Engineering, University of Texas, Austin, July 1991.

[4] K. M. Passino and P. J. Antsaklis. On the optimal control of discrete event systems. In *Proc. 28th IEEE Conf. on Decision and Control*, pages 2713–2718, Tampa, FL, December 1989.

[5] P. J. Ramadge and W. M. Wonham. Supervisory control of a class of discrete event systems. *SIAM J. Control and Optimization*, 25(1):206–230, January 1987.

[6] R. Sengupta and S. Lafortune. A graph-theoretic optimal control problem for terminating discrete event processes. *Discrete Event Dynamic Systems: Theory and Applications*, 2(2):139–172, September 1992.

CHAPTER III

THE WORKSHOP EXERCISE

THE WORKSHOP EXERCISE:
AN INTRODUCTION

During the workshop, there was a special session devoted to one particular problem (namely the cat and mouse control problem, as introduced by Ramadge and Wonham in [1]) in order to give a way of comparing different approaches in the field of discrete event systems.

Although the exercise is just a toy example, the organizers hope that it will contribute in the discussions and shows pro's and cons of each of the approaches dealing with this problem.

The following sections explain the problem in a more detailed way and present briefly the original formulation and solution [1]. In the following papers of the proceedings, different approaches are used to find a solution.

1. THE CAT AND MOUSE MAZE

A cat and a mouse are placed in a maze (see figure 1), with initially the cat in room 2 and the mouse in room 4. Each doorway is denoted by m_i and c_i and can be traversed exclusively by the mouse and by the cat, respectively, in the direction indicated. Each door (with the exception of c_7) can be opened or closed by means of control actions. A controller observes only discrete events generated by sensors in the doors indicating that an animal is just running through.

The control problem is to find a feedback controller such that the closed-loop system satisfies the following two constraints:

1. The animals never occupy the same room simultaneously. If so, the cat will eat the mouse.

2. It is always possible for the cat and the mouse to return back to their initial positions (*i.e.* to the state in which the cat is in room 2 and the mouse in room 4).

If such solution exists then the controller should enable the animals to behave as freely as possible with respect to the given constraints.

2. SOLUTION IN SUPERVISORY CONTROL

The possibilities of movement of the cat and the mouse through the maze are described in [1] using the languages generated by automata $\mathcal{G}_c$ and $\mathcal{G}_m$ (figures 2 and 3). The state spaces of these automata are equal to the set of names of rooms. The transitions are labelled by the names of doors which can be traversed by the cat or by the mouse, respectively. All states are final states, therefore the languages generated are prefix closed, i.e. they are equal to the set of all prefixes of all their words.

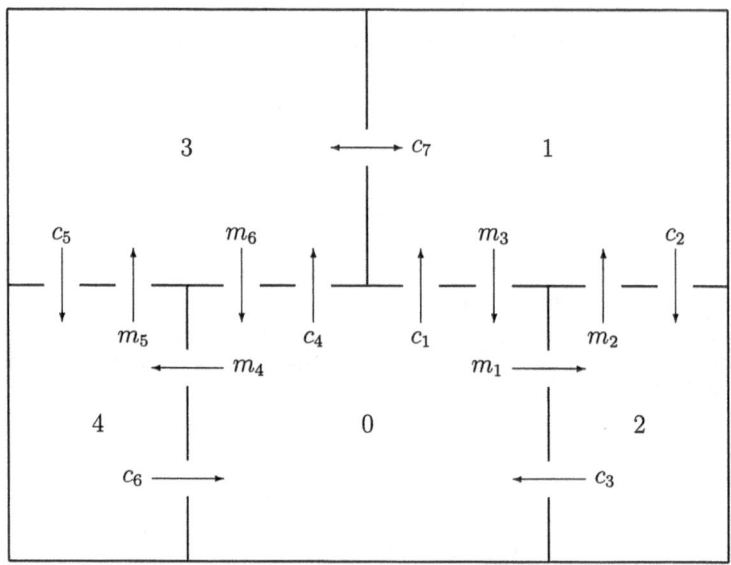

Figure 1: Diagram for the cat and mouse maze

The language for the joint model of cat and mouse is constructed out of the separate automata for cat and mouse using the shuffle product. The state space of the resulting automaton is equal to the Cartesian product of the state space of $\mathcal{G}_c$ and $\mathcal{G}_m$. The states in this new automaton $\mathcal{G}$ are denoted by pairs (i, j), where i is a state of the cat-automaton and j is a state of the mouse-automaton. Transitions in this automaton are of the form:

$$(i, j) \xrightarrow{c_k} (i', j)$$

if there is a transition labelled c_k from i to i' in the automaton $\mathcal{G}_c$, or:

$$(i, j) \xrightarrow{m_k} (i, j')$$

if there is a transition labelled m_k from j to j' in the automaton $\mathcal{G}_m$.

The initial state is $(2, 4)$ and again all states are final states. The transitions are labelled by the set of all door names. By Σ we denote the set of all door names and by $L(\mathcal{G})$ we denote the language generated by $\mathcal{G}$. The language $L(\mathcal{G})$ represents the possible behaviour of the maze if it is not controlled.

The control is defined using the concept of a controlled generator [2] and a set of controllable events $\Sigma_c = \Sigma \setminus \{c_7\}$, where $\setminus$ is the set difference. The set of uncontrollable events is denoted by $\Sigma_u = \{c_7\}$. It is interpreted such that all doors except c_7 can be closed or opened by means of control. The concept of the complete supervisor [2] is playing here the role of feedback interconnection.

Let us consider any language L' which is a subset of the uncontrolled behaviour $L(\mathcal{G})$. It is proved that there exists a supervisor such that the closed-loop behaviour of the system is equal to $\overline{L'}$ ($\overline{L'}$ is the set of all prefixes of L') if and only

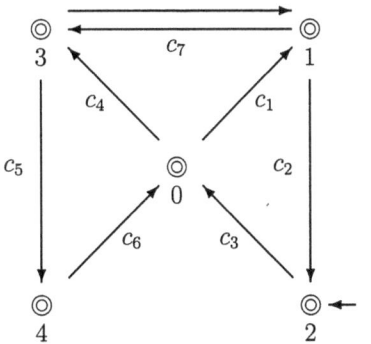
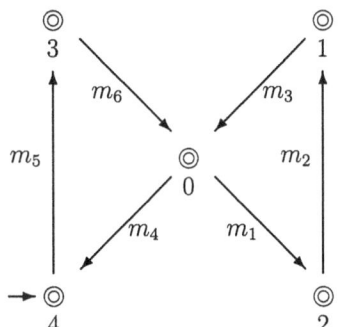

Figure 2: Automaton $\mathcal{G}_c$ for cat-walks

Figure 3: Automaton $\mathcal{G}_m$ for mouse-walks

if L' is the controllable language with respect to $L(\mathcal{G})$ and the set of controllable events Σ_c. Also a construction of this supervisor is presented [1, 2]. On the basis of these results, the control problems can be formulated as problems of finding a controllable language satisfying given constraints.

The maze problem can be formulated using this framework as follows. Let

$$L_m(\mathcal{G}) \;=\; \{w \in L(\mathcal{G}) : \xi((2,4), w) = (2,4)\}$$

where ξ is the transition mapping of $\mathcal{G}$. The language $L_m(\mathcal{G})$ (called the marked language) consist of all words which take the automaton from the initial state to the initial state. It means that it consist of words describing all possible walks of animals ending again at their own homes. This language is recognized by an automaton having all parameters the same as $\mathcal{G}$ but with the set of final states equal to $\{(2,4)\}$. Let us denote this automaton by $\mathcal{G}'$. The language $L_m(\mathcal{G})$ satisfies Constraint 2 of the problem.

To construct an automaton generating a legal language L_g satisfying both constraints, it is possible to remove all states (i, i) from $\mathcal{G}'$. Note that $L_g \subseteq L_m(\mathcal{G})$.

The control problem is to find a language $K \subseteq L_g$ such that K is a controllable language with respect to L and Σ_c, i.e.

$$\overline{K}\Sigma_u \cap L(\mathcal{G}) \subseteq \overline{K}$$

and K is $L_m(\mathcal{G})$-closed, i.e.

$$K \;=\; \overline{K} \cap L_m(\mathcal{G}) \,,$$

where $\overline{K}$ is the set of all prefixes of K. The resulting closed-loop behaviour is then equal to $\overline{K}$.

In [1], an effective procedure for computation of K is given. Moreover the resulting language is the largest one with respect to the set inclusion. This language is called supremal controllable sublanguage of L_g. This solution is the minimally restrictive one as it enables the animals to behave as freely as possible. The result of the computation can be found in figure 4.

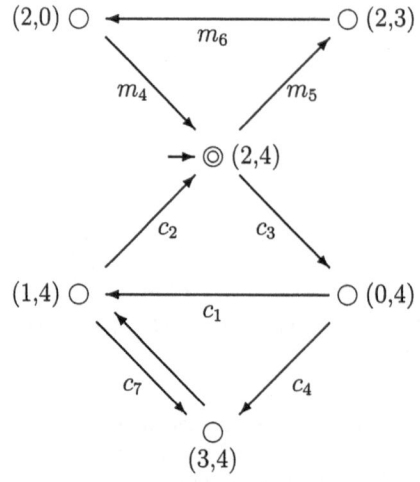

Control Strategy:

- The events m_5 and c_3' are disabled at the state 1.

- Otherwise, all events are enabled at the states 0 and 1.

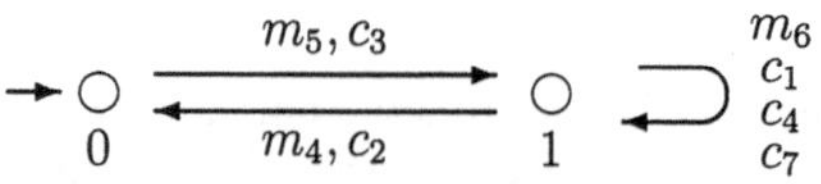

Figure 4: Automaton recognizing the supremal controllable sublanguage of L_g (state (2,4) is the only final state)

Figure 5: Supervisor for cat and mouse

Using theory from [2] the quotient supervisor can be computed which controls the maze such that the closed-loop behaviour is the set of all prefixes of the supremal controllable sublanguage. This supervisor can be found in figure 5.

The control strategy can be summarized as follows: starting with both animals in their original positions, both have the opportunity to move to a new room, but once an animal has moved, the other should stay in its initial room until the other animal has returned back to its initial room. So the cat can do a walk (and the mouse stays home) or the mouse can do a walk (and the cat should stay home). Note that this model does not suppose the simultaneous jumps of animals through doors.

REFERENCES

[1] W.M. Wonham and P.J. Ramadge. On the supremal controllable sublanguage of a given language. *SIAM J. Control Optim.*, 25(3):637–659, May 1987.

[2] P.J. Ramadge and W.M. Wonham. Supervisory control of a class of discrete event processes. *SIAM J. Control Optim.*, 25(1):206–230, January 1987.

THE WORKSHOP EXERCISE USING A TRACE THEORY BASED SETTING

Rein Smedinga*

Abstract. Discrete event systems can be modelled using a trace theory based setting. In this setting the workshop exercise is modelled and solved. To be able to compute the needed systems, a state space model is introduced.

1. INTRODUCTION

For a description of the cat and mouse problem we refer to the introduction in these proceedings. We refer to [1] for the definitions and properties of a logical discrete event system in a trace theory based setting.

Before we start formulating the cat and mouse maze in our setting, we first introduce the state space form.

2. STATE GRAPHS

Definition 1 *A state graph G is defined by:*

$$G = (A, Q, \delta, q, B, T)$$

	↘	*initial state*
with:	●	*non-state* $\notin B \cup T$
A — *the alphabet,*	○	*behaviour state* $\in B \setminus T$
Q — *a (non-empty) set of states,*	◉	*task state* $\in T \setminus B$
$\delta: Q \times A \to Q$ — *the transition function,*	◎	*behaviour/*
$q \in Q$ — *the initial state,*		*task state* $\in B \cup T$
$B \subseteq Q$ — *the behaviour states,*	$\xrightarrow{a}$	*transition labelled a*
$T \subseteq Q$ — *the task states.*		

We assume that δ is a totally defined function. A behaviour state is displayed by an open circle, a non-behaviour state by a solid circle. A larger concentric circle denotes a task state. This leads to the four combinations of possible states.

Each path in such a graph, starting in the initial state and ending in a behaviour state is a behaviour of the system it represents and each path starting in the initial state and ending in a task state is a task of the system.

A state graph is an extension of a finite state automaton in two ways: first, it contains two kinds of final states (behaviour states and task states) and, second, the number of states need not be finite[1].

---

*Department of computing science, University of Groningen, P.O.Box 800, NL-9700 AV Groningen, the Netherlands, e-mail: rein@cs.rug.nl

[1]Although we will only use state graphs for computations if the number of states is finite.

Example 1 The system

$$P = \langle \{a, b\}$$
$$, \{\epsilon, ab, aa, aba\}$$
$$, \{ab\} \rangle$$

can be displayed using the
graph at the right. □

3. MODELLING THE CAT AND MOUSE MAZE

In our trace theory based setting (see [1]) the cat and mouse maze can be modelled
by giving all events that are possible (i.e., $c_1, \ldots, c_7, m_1, \ldots, m_6$) and all possible
behaviour of the system. Each behaviour that results in the cat and mouse being
back in their original positions is considered a completed task.

The formulation of the control problem means dividing the events in controllable
and uncontrollable ones. Only c_7 is uncontrollable. In order to formulate the
control goal in our setting we add one extra event, namely x, denoting that the
cat eats the mouse. The uncontrolled exogenous behaviour of the system now
equals

$$\{\epsilon, x, c_7, c_7 x, c_7 c_7, c_7 c_7 x, \ldots\} = \mathbf{pref}(c_7^* x)$$

It is not difficult to find a model in state space for this system. We number the
possible states by (ab) meaning that the cat is in room a and the mouse is in
room b. We add one more state, namely †, to denote the situation when the cat
has eaten the mouse. Transitions in the modelling graph are (a, b, a', and b'
represent one of $0, \ldots, 4$):

$$(aa) \xrightarrow{x} \dagger \qquad \text{cat eats mouse}$$
$$(ab) \xrightarrow{y} (a'b) \quad \text{cat goes from } a \text{ to } a' \text{ via door } y \qquad y = c_1, \ldots, c_7$$
$$(ab) \xrightarrow{y} (ab') \quad \text{mouse goes from } b \text{ to } b' \text{ via door } y \quad y = m_1, \ldots, m_6$$

Initially the cat is in room 2 and the mouse is in room 4. So (24) is the initial
state. We let cat and mouse freely walk through the maze but only have completed
behaviour (tasks) if cat and mouse have returned to their initial positions. Also
if the cat has eaten the mouse we say we have a completed task. So the states
denoted by (24) and † are task states. In the diagram on the following page all
transitions are given.[2]

Our control problem now is finding some controller that forbids cat and mouse
to go through some doors if this leads to the mouse being eaten by the cat. In
our framework this means that we are looking for a controller that allows in each
situation those events that do no lead to disaster.

First, suppose event x is the only uncontrollable event. We choose:

$$L_1 := L_{min} = L_{max} = \langle \{x\}, \epsilon \rangle$$

i.e., event x may not occur. The remaining step is to compute $\mathbf{des}(F(P, L_1))$.

---

[2]But all transitions going to the only non-state and that non-state itself are not drawn.

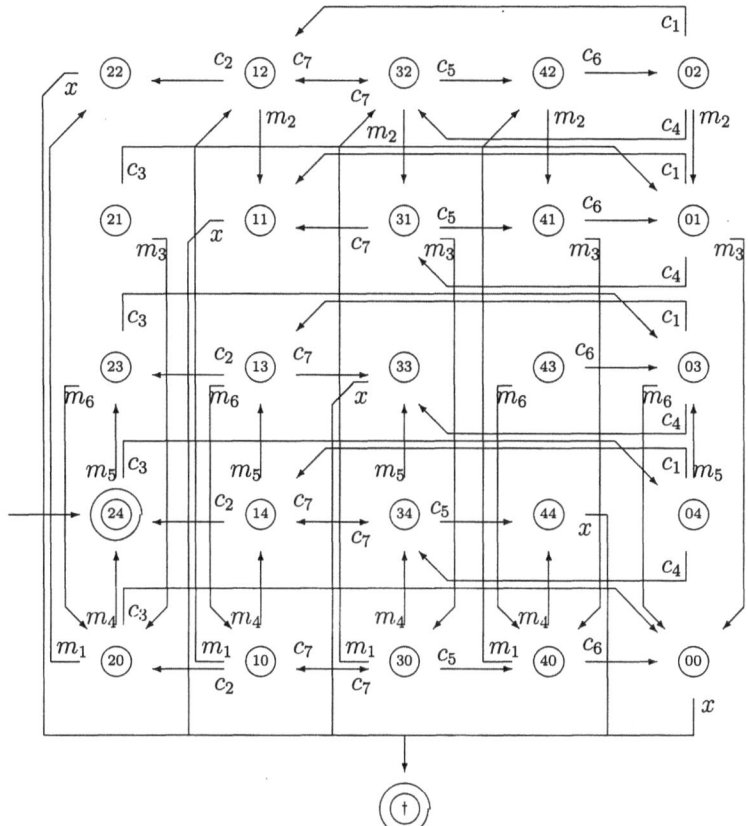

4. COMPUTATIONS WITH STATE GRAPHS

In order to compute controllers effectively, we have to define how the operators on discrete event systems can be applied on state graphs. In this section we only give a brief outline of these operations on graphs. Moreover, we only consider finite state graphs here and will not give proofs. The reader is referred to [2] and [3] for this.

Suppose system P is represented by the graph G_P. From this graph we can easily construct a graph for $\sim P$ by simply reversing the meaning of each of the states, i.e., a behaviour state becomes a non-behaviour state, a task-state a non-task state, etc.

From G_P we construct a graph for $P \lceil A$ by replacing every transition not labelled with an event from A by an ϵ-transition. The so constructed non-deterministic graph can be made deterministic again by applying (extended) standard techniques (see [4]).

A graph for $\mathbf{des}(P)$ can be derived from G_P by adding one fresh non-state to the state set Q and redirecting all transitions going to a non-behaviour state to

in P	in $\sim P$
●	◎
○	◉
◉	○
◎	●

this new non-state. Afterwards all states that cannot be reached any more can be removed.

Given P and R in graph-form, the connection can be modelled using the cartesian product of these two graphs, where a state in the product graph is a behaviour state (or task state) only if both the corresponding states in the original graphs are behaviour state (or task state, respec-tively). Transitions in the product graph are according to the scheme above.

$a \in$	P	R	$P \parallel R$
$aP \cap aR$	$\boxed{1} \xrightarrow{a} \boxed{2}$	$\boxed{3} \xrightarrow{a} \boxed{4}$	$\boxed{13} \xrightarrow{a} \boxed{24}$
$aP \setminus aR$	$\boxed{1} \xrightarrow{a} \boxed{2}$	$\boxed{3}$	$\boxed{13} \xrightarrow{a} \boxed{23}$
$aR \setminus aP$	$\boxed{1}$	$\boxed{3} \xrightarrow{a} \boxed{4}$	$\boxed{13} \xrightarrow{a} \boxed{14}$

($\boxed{}$ stands for an arbitrary state)

The external connection can now be performed on state graphs by first com-puting the normal connection and then restricting the system to the appropriate alphabet.

5.   THE FIRST CONTROLLER

Computing $\mathbf{des}(F(P, L_1))$ leads to the connection $P \parallel \mathbf{des}(F(P, L_1))$ as given below. Note that event x never occurs in the connection: $P \lceil \mathbf{des}(F(P, L_1)) = L_1$.

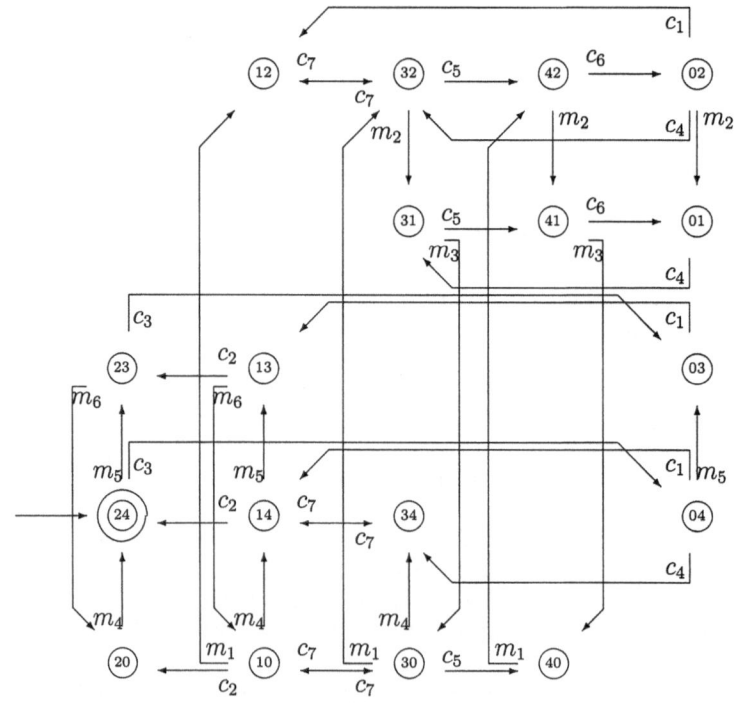

We now consider the situation where also event c_7 is uncontrollable. Then we have

$$L_2 := \langle \{x, c_7\}, c_7^* \rangle$$

I.e., c_7 may occur as often as possible, but x may never occur.

Computing the controller $R = \mathbf{des}(F(P, L_2))$ leads to the connection of system and controller given on the right. Although we indeed have established the minmax condition, i.e., $P \,]\![\, R = L_2$, the connection itself $P \parallel R$ may deadlock in state (03). So we should find some other controller that satisfies the minmax condition and also leads to a connection that is free of lock.

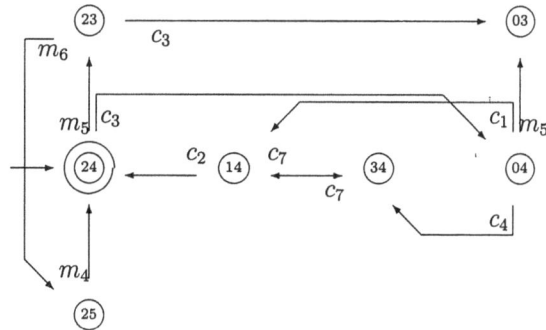

We use the method described in [1] and compute the greatest subsystem of the controller $\mathbf{des}(F(P, L_2))$ that is free of lock in connection with P, so we compute $\Lambda(P, \mathbf{des}(F(P, L_2)))$. Before giving the results, we explain how these computations can be done on state graphs.

6. COMPUTATIONS WITH STATE GRAPHS (part 2)

Locked traces in a graph can be found by applying the following algorithm:

First, reverse all transitions in the graph; second, collect all states reachable from a task state in this reversed graph (the remaining states are called locked states), and third, replace in the original graph all states by non-states and replace each locked state by a behaviour/task state

After this algorithm the new graph represents the system $\langle aP, T, T \rangle$ where T is the set of locked traces.

At last, we need to explain how $P \setminus T$ can be computed using state graphs. Again we use a cartesian product of the graphs of P and T, but this time a state in the product graph is a behaviour state if the corresponding state in the graph of P is a behaviour state and the corresponding state in the graph of T is not (and similar for task states). Transitions are now defined according to the scheme at the right.[3]

P	T	$P \setminus T$
$\boxed{1} \xrightarrow{a} ②$	$\boxed{3} \xrightarrow{a} ④$	$\boxed{13} \xrightarrow{a} \bullet$
$\boxed{1} \xrightarrow{a} ②$	$\boxed{3} \xrightarrow{a} \bullet$	$\boxed{13} \xrightarrow{a} \circ$
$\boxed{1} \xrightarrow{a} ②$	$\boxed{3} \xrightarrow{a} ⊙$	$\boxed{13} \xrightarrow{a} \circ$
$\boxed{1} \xrightarrow{a} ②$	$\boxed{3} \xrightarrow{a} ④$	$\boxed{13} \xrightarrow{a} \bullet$

---

[3]Only the case of state 2 being a behaviour state is given, but a similar scheme can be made if 2 is a task state or a behaviour/task state. If 2 is a non-state the corresponding state in $P \setminus T$ is always a non-state.

7.   COMPUTING THE SECOND CONTROLLER

Although it is clear from the graph of $P \parallel \mathbf{des}(F(P,L))$ that state (03) is a deadlock state, we give the graph of $\mathbf{lock}(P \parallel \mathbf{des}(F(P,L)))$ at the right. The next step is to compute iteratively the chain

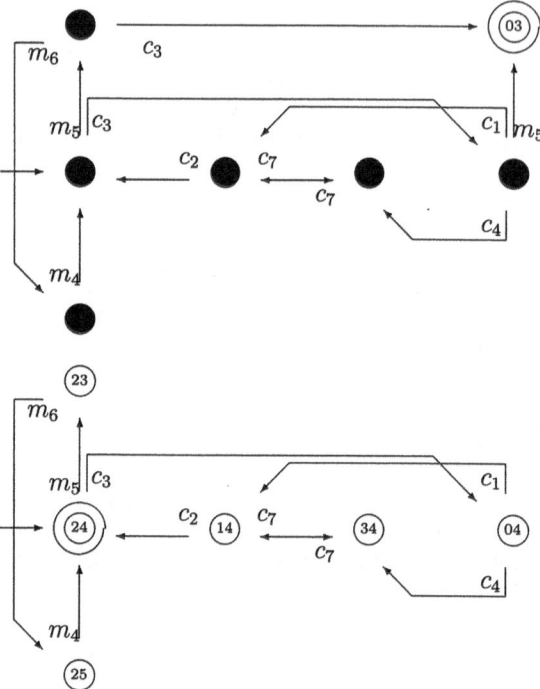

$$R_0 = \mathbf{des}(F(P,L))$$
$$R_i = R_{i-1} \setminus \mathbf{lock}(P, R_{i-1})$$

which reaches its fix-point after one step.
The second graph at the right shows the resulting connection $P \parallel \Lambda(P, \mathbf{des}(F(P,L)))$, that indeed is free of lock and has the desired behaviour.
The solution, as found here, is equivalent to the one Ramadge and Wonham found in [5]: the only possibilities are to let the mouse do a run or to let the cat do a run.
Only if the walking animal has returned in its initial room, both animals are allowed to do a first step. As soon as one of the animals starts walking the other is no longer allowed to leave its room.

REFERENCES

[1] R. Smedinga. An overview of results in discrete event systems using a trace theory based setting. *In this volume*, pages 43–56.

[2] R. Smedinga. The reflection operator in discrete event systems. Technical Report CS9201, Department of computing science, University of Groningen, 1992.

[3] R. Smedinga. Discrete event systems. course-notes, Department of computing science, University of Groningen, 1992.

[4] J.E. Hopcroft and J.D. Ullman. *Introduction to automata theory, languages, and computation.* Addison Wesley, 1979.

[5] W.M. Wonham and P.J. Ramadge. On the supremal controllable sublanguage of a given language. *SIAM journal on control and optimalisation*, 25 (3), 1987.

A PETRI NETS-BASED APPROACH TO THE
MAZE PROBLEM SOLVING

František Čapkovič*

Abstract. The cat-and-mouse problem formulated by Ramadge and Wonham in [1] is solved in this paper by means of Petri nets. The behaviour of the animals is expressed in the form of the abstract linear discrete dynamical systems. A simple procedure of the control synthesis is presented. It compares (step-by-step) the possibilities of the further development of the animals behaviour with a knowledge base expressing the predefined constraints imposed on their behaviour.

1. INTRODUCTION

The DEDS are often described by different types of the Petri nets (PN). Let us go to illustrate the PN-based approach to the DEDS modelling and control on the solving of the maze problem presented in [1]. The PN are understood here to be the directed bipartite graphs. More details about such an understanding the PN can be found in [2], where the corresponding formalism is presented. Consequently, the following linear discrete dynamic model of the DEDS can be written

$$x_{K+1} = x_K + B.u_K \qquad (1)$$
$$B = G^T - F \qquad (2)$$
$$F.u_K \leq x_K \qquad (3)$$

where $K = 0, N$; is the discrete step of the DEDS dynamics development

x_K is the n-dimensional state vector of the system in the step K

u_K is the m-dimensional control vector of the system in the step K

B, F, G are respectively, $(n \times m)$, $(n \times m)$ and $(m \times n)$-dimensional structural matrices of constant elements

T symbolises the matrix or vector transposition

The matrices F, G are the arcs incidence matrices. However, they influence the PN dynamics (moving tokens among the PN positions) too. Their elements acquire the values from the set $\{0, 1\}$. The matrix F describes the arcs oriented from the PN positions to the transitions (the PN tokens leave throughout them the positions, therefore F must be taken with the minus sign). The matrix G describes the arcs oriented from the transitions to the positions (the tokens flow throughout them into the positions, therefore G must be taken with positive

*Institute of Control Theory and Robotics, Slovak Academy of Sciences, Dúbravská cesta 9, 842 37 Bratislava, ČSFR

sign). The components of the state vector express the states of the elementary subprocesses of the DEDS - the activity is symbolised by means of the presence of the token (when 1); the passivity is expressed by means of the absence of the token (when 0). The components of the control vector represent the elementary discrete events in the DEDS - the appearance of an event is symbolised by means of the open transition (when 1); the absence of it is expressed like the closed transition (when 0).

2. POSSIBILITIES OF THE DEDS CONTROL

In order to find the suitable control vector u_K we shall use the following procedure

$$\overline{x}_K = \underline{\text{neg}}\, x_K = \mathbf{1}_n - x_K \tag{4}$$

$$v_K = F^T \underline{\text{and}}\, \overline{x}_K \tag{5}$$

$$w_K = \underline{\text{neg}}\, v_K = \mathbf{1}_m - v_K = \mathbf{1}_m - (F^T \underline{\text{and}}\, (\mathbf{1}_n - x_K)) \tag{6}$$

where $\underline{\text{neg}}$ is the operator of logical negation

$\underline{\text{and}}$ is the operator of logical multiplying

$\mathbf{1}_n$ is n-dimensional constant vector with elements equaled to 1

v_K, w_K are, respectively, m-dimensional auxiliary vector and m-dimensional vector of the base for the control vector choice

The vector w_K expresses the possible candidates for the control vector u_K in the step K. When only one of its components is different from zero, it can be used to be the control vector, i.e. $u_K = w_K$. When there are several components of the w_K different from zero, the control vector u_K has to be chosen on the basis of additional information about the actual control task. The choice of the control vector can be made either by an expert in the corresponding domain or automatically on the basis of the rules predefined by means of the constraints imposed upon the task in question. The prescribed conditions which must be satisfied during the actual task solving (including the desired final aim of the control process in case of control synthesis) are understood to be such constraints. Knowledge about the problem to be solved is expressed in the form of the problem oriented knowledge base. More details about the knowledge representation can be found in [4]. Information about the possibilities of the future development of the system can be given by the following K-variant discrete dynamical system

$$x_{K+1} = A_K.x_K, \quad K = 0, N \tag{7}$$

$$A_K = A_K(u_K) \quad ; \quad A_K \subseteq (F.G)^T \tag{8}$$

where A_K is the $(n \times n)$-dimensional K-variant matrix given on the basis of the $(n \times n)$-dimensional constant matrix $(F.G)^T$ expressing the apportioning the PN transitions to the oriented arcs among the PN positions. Its elements express the actual states of the elementary transitions (i.e. the elementary discrete events given in (1)–(3) by means of the vector u_K). Its dependency on the u_K is symbolised like $A_K(u_K)$. As to its creation, more details can be found in [3]. By means of this matrix we can write (see [3])

$$x_K = \underbrace{A_{K-1}.A_{K-2}.\ldots.A_1.A_0}_{\Phi_{K,0}}.x_0 = \Phi_{K,0}.x_0 \tag{9}$$

$$\Phi_{K,M} = A_{K-1}.A_{K-2}.\ldots.A_{M+1}.A_M \quad , \quad \Phi_{K,K} = I_n \tag{10}$$

where $\Phi_{K,M}$, $K > M$ is, in general, the $(n \times n)$-dimensional transition matrix expressing the possible trajectories of the system in case of its transition from the state x_M in the step $M < K$ to the actual state x_K in the step K. The matrix I_n is the $(n \times n)$-dimensional identity matrix. The rough procedure of the control synthesis can be verbally expressed as follows

- START

- $K = 0$ i.e. $x_K = x_0$

LABEL:

- generation of the control base w_K

- generation of the possible control vectors $u_K \in w_K$

- generation of the corresponding model responses x_{K+1}

- confrontation with the knowledge base

- choice of a suitable control possibility

- *if* (the expected solution was found) *then* (*goto* END) *else* (*begin* $K = K + 1$; *goto* LABEL; *end*)

- END

The expected solution of a problem (e.g. the expected comeback of both animals in the example given in [1] to their initial rooms) represents the final aim of the control process. It is described as a part of the knowledge base as well as the rules how to satisfy all of the prescribed restrictions (e.g. avoiding the crash, as free behaviour of the animals as possible, etc.).

3.   THE PETRI NETS-BASED MODEL OF THE MAZE

Consider the rooms of the maze to be the Petri nets positions (places) and the doorways of the animals to be the Petri nets transitions. The corresponding PN-based representation of the maze is given on figure 1. It can be seen that the uncontrolled door c_7 is replaced by two PN transitions c_7, c_8 in order to avoid the problems specific for the PN. Both the cat and the mouse can be expressed, as to their dynamic behaviour, by means of the linear discrete dynamic model (1)–(3). The parameters of the cat model are the following

$$
\begin{aligned}
n &= 5 \\
m_c &= 8
\end{aligned}
\quad
F_c =
\begin{pmatrix}
1 & 0 & 0 & 1 & 0 & 0 & 0 & 0 \\
0 & 1 & 0 & 0 & 0 & 0 & 1 & 0 \\
0 & 0 & 1 & 0 & 0 & 0 & 0 & 0 \\
0 & 0 & 0 & 0 & 1 & 0 & 0 & 1 \\
0 & 0 & 0 & 0 & 0 & 1 & 0 & 0
\end{pmatrix}
\quad
G_c^T =
\begin{pmatrix}
0 & 0 & 1 & 0 & 0 & 1 & 0 & 0 \\
1 & 0 & 0 & 0 & 0 & 0 & 0 & 1 \\
0 & 1 & 0 & 0 & 0 & 0 & 0 & 0 \\
0 & 0 & 0 & 1 & 0 & 0 & 1 & 0 \\
0 & 0 & 0 & 0 & 1 & 0 & 0 & 0
\end{pmatrix}
$$

and the parameters of the mouse model are

$$n = 5, \ m_m = 6, \ F_m = \begin{pmatrix} 1 & 0 & 0 & 1 & 0 & 0 \\ 0 & 0 & 1 & 0 & 0 & 0 \\ 0 & 1 & 0 & 0 & 0 & 0 \\ 0 & 0 & 0 & 0 & 0 & 1 \\ 0 & 0 & 0 & 0 & 1 & 0 \end{pmatrix} \quad G_m^T = \begin{pmatrix} 0 & 0 & 1 & 0 & 0 & 1 \\ 0 & 1 & 0 & 0 & 0 & 0 \\ 1 & 0 & 0 & 0 & 0 & 0 \\ 0 & 0 & 0 & 0 & 1 & 0 \\ 0 & 0 & 0 & 1 & 0 & 0 \end{pmatrix}$$

The matrix A and the corresponding matrix A_K in both of the previous cases are given as follows

$$A_c = \begin{pmatrix} 0 & 0 & 1 & 0 & 1 \\ 1 & 0 & 0 & 1 & 0 \\ 0 & 1 & 0 & 0 & 0 \\ 1 & 1 & 0 & 0 & 0 \\ 0 & 0 & 0 & 1 & 0 \end{pmatrix} \quad {}^c A_K = \begin{pmatrix} 0 & 0 & c_3^K & 0 & c_6^K \\ c_1^K & 0 & 0 & c_8^K & 0 \\ 0 & c_2^K & 0 & 0 & 0 \\ c_4^K & c_7^K & 0 & 0 & 0 \\ 0 & 0 & 0 & c_5^K & 0 \end{pmatrix} \quad \begin{array}{l} K = 0, N; \\ c_i^K \in \{0,1\} \\ i = 1,8 \end{array}$$

$$A_m = \begin{pmatrix} 0 & 1 & 0 & 1 & 0 \\ 0 & 0 & 1 & 0 & 0 \\ 1 & 0 & 0 & 0 & 0 \\ 0 & 0 & 0 & 0 & 1 \\ 1 & 0 & 0 & 0 & 0 \end{pmatrix} \quad {}^m A_K = \begin{pmatrix} 0 & m_3^K & 0 & m_6^K & 0 \\ 0 & 0 & m_2^K & 0 & 0 \\ m_1^K & 0 & 0 & 0 & 0 \\ 0 & 0 & 0 & 0 & m_5^K \\ m_4^K & 0 & 0 & 0 & 0 \end{pmatrix} \quad \begin{array}{l} K = 0, N; \\ m_i^K \in \{0,1\} \\ i = 1,6 \end{array}$$

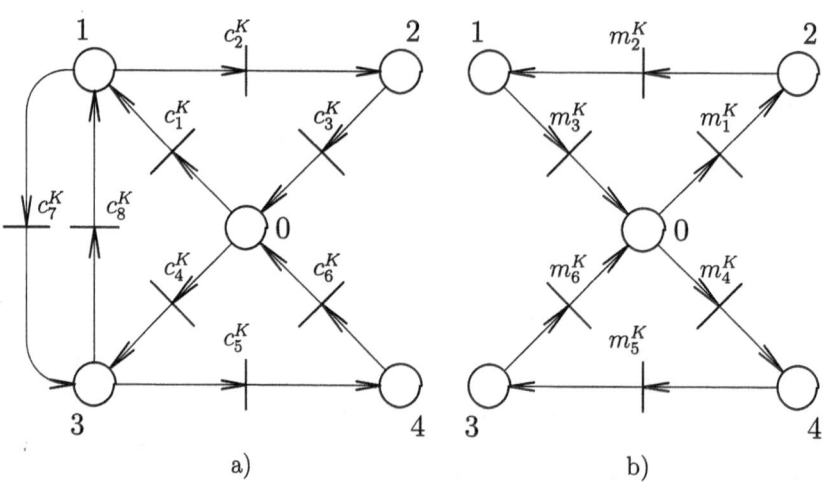

Figure 1: The PN-based representation of the maze for the cat and mouse. a) possibilities of the cat behaviour b) possibilities of the mouse behaviour

The initial state vectors of the cat and the mouse are:

$$^{c}x_0 = (0\,0\,1\,0\,0) \quad , \quad ^{m}x_0 = (0\,0\,0\,0\,1)^T$$

The structure of the control vectors of the animals is:

$$
\begin{aligned}
^{c}u_K &= (c_1^K, c_2^K, c_3^K, c_4^K, c_5^K, c_6^K, c_7^K, c_8^K)^T \,, \quad c_i^K \in \{0,1\}, \quad i = 1,8 \\
^{m}u_K &= (m_1^K, m_2^K, m_3^K, m_4^K, m_5^K, m_6^K)^T \,, \quad m_i^K \in \{0,1\}, \quad i = 1,6
\end{aligned}
$$

4. THE CONSTRAINTS ACCEPTATION

As to the first constraint (the animals must never occupy the same room simultaneously), we can see for example that it is a risk to open the door m_5 for the mouse being in the room 4 when the cat is in the room 1, even, the risk is double when the door c_2 is closed. Analogically, it is a risk to open the door m_2 for the mouse being in the room 2 when the cat is in the room 3, especially when the door c_5 is closed. On the other hand, we can also see that simultaneous opening the following pairs of doors (c_3, m_3), (c_3, m_6), (c_6, m_3), (c_6, m_6), (m_4, c_5), (c_4, m_5), (m_1, c_2), (c_1, m_2) is impossible because of those demand. We feel that it is dangerous for the mouse being in the room 0 to use the door m_1, or the door m_4 when the cat is in the room 1 or in the room 3, because it leads to a crash in the next or further step. However, the conflict of interests of the animals can also be pointed out exactly by means of the relation

$$^{c}\Phi_{K,0}^T \cdot {}^{m}\Phi_{K,0} = 0, \quad K = 1, N \tag{11}$$

It shows us which trajectories should be avoided. In order to satisfy the second demand (the cat and mouse should always have a possibility to return to their initial position) a support of this should be made in such a way that the doors making the satisfying the demand possible should be opened and the doors of the adversary restraining this should be closed. The constraints help us to choose the best control vector u_K for both animals when the vector w_K gives us more possibilities. The mentioned policies can be expressed in the form of rule-based knowledge. When the knowledge base is sufficient the process of the automatic synthesis of the control can be obtained.

5. THE STEP-BY-STEP PROBLEM SOLVING

As a result of the PN-based approach (1)–(7) with respect to the knowledge base mentioned above the behaviour of the animals is described in detail in the Tab. 1 and graphically illustrated on the figure 2.

As we can see in the table, the uncontrollable doors c_7, c_8 become evident in the step $K = 2$. We cannot prohibit to the cat using those doors and, consequently, the doors must be respected. What we can only do in order to avoid a crash of the animals in the next step is, that we can support opening the door c_2 if the cat is in the room 1 (this door should be opened because of the possibility of its returning to the initial state - i.e to the room 2). In addition to this fact, opening the door c_2 has priority to opening the mouse door m_1 leading the mouse to the room 2 what is undesirable. We prohibit this mouse door, in order to avoid a possible crash, and support opening the door m_4 leading the mouse to their initial state - the room 4.

PROCESS DEVELOPMENT		
Step	**Cat Behaviour**	**Mouse Behaviour**
$K = 0$	$^c x_0 = (0\,0\,1\,0\,0)^T$ $^c \bar{x}_0 = (1\,1\,0\,1\,1)^T$ $^c w_0 = (0\,0\,1\,0\,0\,0\,0\,0)^T$ $^c u_0 = {}^c w_0$ $^c x_1 = {}^c x_0 + B_c.{}^c u_0$ $^c x_1 = (1\,0\,0\,0\,0)^T$	$^m x_0 = (0\,0\,0\,0\,1)^T$ $^m \bar{x}_0 = (1\,1\,1\,1\,0)^T$ $^m w_0 = (0\,0\,0\,0\,1\,0)^T$ $^m u_0 = {}^m w_0$ $^m x_1 = {}^m x_0 + B_m.{}^m u_0$ $^m x_1 = (0\,0\,0\,1\,0)^T$
	$^c x_1 \neq {}^m x_1$	
$K = 1$	$^c w_1 = (1\,0\,0\,1\,0\,0\,0\,0)^T$	$^m w_1 = (0\,0\,0\,0\,0\,1)^T$
	the control possibilities are: $\{\underline{c_1},\ \underline{c_4},\ \underline{m_6}\}$ all of them are possible	
	$^c u_1^1 = (1\,0\,0\,0\,0\,0\,0\,0)^T$ $^c x_2^1 = (0\,1\,0\,0\,0)^T$ $^c u_1^2 = (0\,0\,0\,1\,0\,0\,0\,0)^T$ $^c x_2^2 = (0\,0\,0\,1\,0)^T$	$^m u_1 = {}^m w_1$ $^m x_2 = (1\,0\,0\,0\,0)^T$ $=$ n o n e $=$ $=$ n o n e $=$
	$^c x_2^1 \neq {}^m x_2;\ ^c x_2^2 \neq {}^m x_2$	
$K = 2$	$^c w_2^1 = (0\,1\,0\,0\,0\,0\,1\,0)^T$ $^c w_2^2 = (0\,0\,0\,0\,1\,0\,0\,1)^T$	$^m w_2 = (1\,0\,0\,1\,0\,0)^T$ $=$ n o n e $=$
	the control possibilities are: $\{c_2,\ c_7,\ \cancel{m_1},\ \underline{m_4}\};\ c_2$ has priority to m_1 $\{\cancel{c_5},\ c_8,\ \cancel{m_1},\ \underline{m_4}\};\ m_4$ has priority to c_5	
	$^c u_2^{11} = (0\,1\,0\,0\,0\,0\,0\,0)^T$ $^c u_2^{12} = (0\,0\,0\,0\,0\,0\,1\,0)^T$ $^c u_2^2 = (0\,0\,0\,0\,0\,0\,0\,1)^T$ $^c x_3^{11} = (0\,0\,1\,0\,0)^T = {}^c x_0$ $^c x_3^{12} = (0\,0\,0\,1\,0)^T$ $^c x_3^2 = (0\,1\,0\,0\,0)^T$	$^m u_2 = (0\,0\,0\,1\,0\,0)^T$ $=$ n o n e $=$ $=$ n o n e $=$ $^m x_3 = (0\,0\,0\,0\,1)^T = {}^m x_0$ $=$ n o n e $=$ $=$ n o n e $=$

Table 1: The solution of the problem.

The other way round, opening the door m_4 has priority to opening the cat door c_5 leading the cat to the mouse initial state (the room 4). Hence, in the step $K = 2$ we prohibit opening the doors m_1, c_5 (in the table they are struck out) and support opening the doors m_4 and c_2 (in the table they are underlined). The permanently open doors c_7, c_8 must be accepted in the control synthesis task. Consequently, the process of returning the cat into the room 2 can be infinite. The solution of the task can be seen better on figure 2.

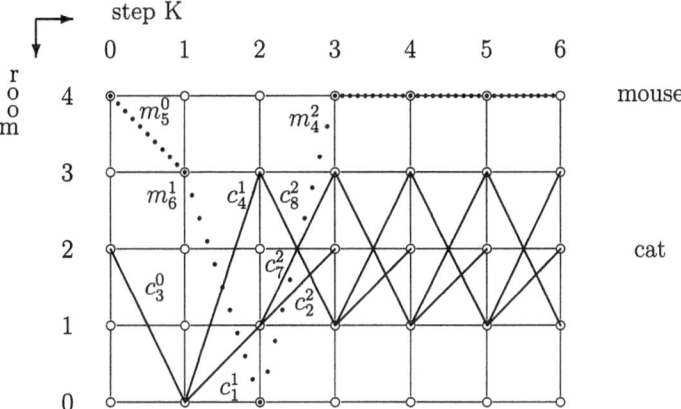

Figure 2: The graphical expression of the task solution.

REFERENCES

[1] Wonham, W.M., Ramadge, P.J.: On the supremal controllable sublanguage of a given language. SIAM J. Control and Optimization. Vol. 25, No 3, May 1987, pp. 637-659

[2] Čapkovič, F.: Petri nets-based computer aided synthesis of control systems for DEDS. In: Barker, H.A. (Ed.): Computer Aided Design in Control Systems. Preprints of the 5-th IFAC Symp., Swansea, U.K., July 1991, Pergamon Press, pp. 409-414

[3] Čapkovič, F.: Systems theory views on systems discrete by nature. In: Vichnevetsky, R. and J.J.H. Miller (Eds.):Proceedings of the 13-th IMACS World Congress on Computation and Applied Mathematics, Dublin, Ireland, Trinity College Press, 1991, pp. 1308-1309

[4] Čapkovič, F.: A representation of rule-based types of reasoning by means of an abstract dynamic system development. Systems Science. Vol. 17, No 4, 1991, pp. 35-54

THE CAT-AND-MOUSE PROBLEM AS A SYSTEM OF BOOLEAN EQUATIONS

Otakar Kříž*

Abstract. The problem is defined in the setting of logical automata and the control of the doors is given by a set of logical functions understood as output function of a Moore automaton describing possible development of the whole system.

The original problem was introduced in [1].

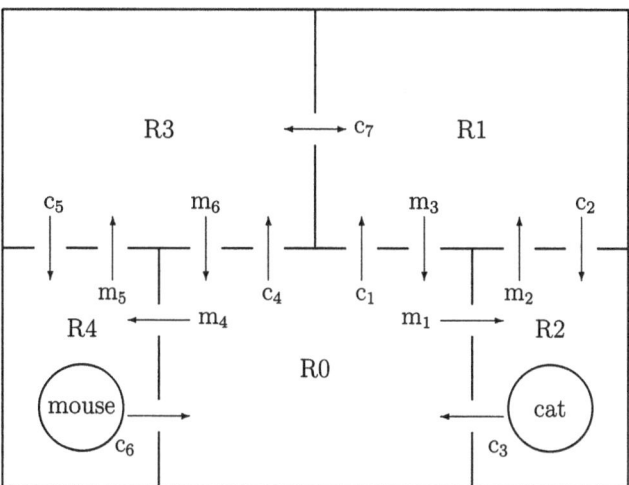

Figure 1: Topology of the maze

The mouse and cat can move in the maze according to figure 1. The control of the doors should guarantee that

1. animals do not occupy the same room simultaneously
2. animals have always possibility to return to their original "home" room (i.e. cat to the room $R2$ and mouse to the $R4$)
3. animals are not restricted in their movement unless it violates preceding conditions

*Institute of Information Theory and Automation, Czechoslovak Academy of Sciences, 182 08 Praha, Czechoslovakia, kriz@cspgas11.bitnet

Note that doors are specific for each animal, they are of one-way type and they are controllable i.e. they can be either closed or opened by the controller. The only exception is the door c_7 that is of two-way type and always opened.

The animals are autonomous subjects in the sense that the controller cannot make an animal pass through a specific door. Therefore, the only way of fulfilling the constraints is via prohibiting the animals to enter a room by closing the respective door.

Let's us denote by $MR0, MR1, MR2, MR3, MR4$ the fact that the mouse is in the room $0, 1, 2, 3, 4$ and similarly, by $CR0, CR1, CR2, CR3, CR4$ presence of the cat in rooms $0, 1, 2, 3, 4$.

In the sequel, we shall need the information that an animal has really passed through a door. Therefore, each door is equipped not only with the door-closing element but also with an indicator. We shall suppose existence of two types of signals for each door marked with prefixes $\$-$ and $\#-$ to distinguish them from door-closing commands that will stay without prefixes. E.g. $\$m_5$ (i.e. $\$m_5 = 1$) expresses the fact that the mouse has just passed through the door m_5 from room R_4 to R_3 and logical variable $\#m_5$ carries the information that the mouse is somewhere near the door m_5 with potential intention to pass through but it has not passed yet completely. If we require that the animal is not hurt by the closing door and suppose it may change its mind when entering the aperture of the door and return back even several times, then the technical realization of obtaining such signals may be little complicated. E.g. 4 light-sensitive elements (2 before and 2 after the door) indicate interruption of the light beam when the animal moves about and the required signals $\$m_5, \#m_5$ are outputs of an automaton with 9 states indicating various phases of the animal's passing through the door. But let us return to the original problem whose setting will be given also in terms of an logical automaton:

MOORE AUTOMATON

$$\mathcal{A} = (\mathcal{I}, \mathcal{S}, \mathcal{O}, \delta, \lambda)$$

$$\mathcal{I} = (\$m_1, \$m_2, \ldots, \$c_7, \$m_1 \cdot \$c_1 + \cdots + \$m_6 \cdot \$c_7, \overline{\$m_1} \cdot \overline{\$m_2} + \cdots)$$

$$\mathcal{S} = \{(MR0, MR1, \ldots, MR4) \times (CR0, CR1, \ldots, CR4)\}$$

$$\mathcal{O} = \{\overline{m_1}, m_1\} \times \{\overline{m_2}, m_2\} \times \cdots \times \{\overline{c_6}, c_6\}$$

$$\delta : \mathcal{I} \times \mathcal{S} \to \mathcal{S}$$

$$\lambda : \mathcal{S} \to \mathcal{O}$$

The possible movement of the mouse respecting the topology of the maze is described by the following set of equations

$$NMR0 = MR0 \cdot \overline{\$m_4} \cdot \overline{\$m_1} + MR1 \cdot \$m_3 + MR3 \cdot \$m_6$$
$$NMR1 = MR1 \cdot \overline{\$m_3} + MR2 \cdot \$m_2$$
$$NMR2 = MR2 \cdot \overline{\$m_2} + MR0 \cdot \$m_1 \qquad (1)$$
$$NMR3 = MR3 \cdot \overline{\$m_6} + MR4 \cdot \$m_5$$
$$NMR4 = MR4 \cdot \overline{\$m_5} + MR0 \cdot \$m_4$$

Interpretation of the names of logical functions,that are expressing the mapping δ, is following:

$NMR0$ is equal to logical 1 if the mouse is entering the room 0 in the next step

$\$m_1$ stands for mouse passing through the door m_1

$\overline{\$m_1}$ means mouse is not entering the door m_1

m_1 means the mouse door m_1 from room 0 to room 2 is closed

$\overline{m_1}$ the door m_1 is opened.

In fact, each equation is a sort of continuity equation. Similar functions describe possible behaviour of the cat:

$$NCR0 = CR0 \cdot \overline{\$c_4} \cdot \overline{\$c_1} + CR2 \cdot \$c_3 + CR4 \cdot \$c_6$$
$$NCR1 = CR0 \cdot \$c_1 + CR1 \cdot \overline{\$c_2} \cdot \overline{\$c_7} + CR3 \cdot \$c_7$$
$$NCR2 = CR1 \cdot \$c_2 + CR2 \cdot \overline{\$c_3} \qquad (2)$$
$$NCR3 = CR0 \cdot \$c_4 + CR1 \cdot \$c_7 + CR3 \cdot \overline{\$c_7} \cdot \overline{\$c_5}$$
$$NCR4 = CR3 \cdot \$c_5 + CR4 \cdot \overline{\$c_6}$$

As, in general, both animals might move independently, there exist 25 "double" states given by Cartesian product

$$\{MR0, MR1, MR2, MR3, MR4\} \times \{CR0, CR1, CR2, CR3, CR4\}$$
$$\text{i.e. } \{(MR0, CR0), (MR0, CR1), \ldots (MR4, CR4)\}.$$

In the sequel, the "algebraic form" of state will be used, too, e.g.CR0.MR4 is taken as equivalent to (CR0, MR4). Constraints imposed on the required behaviour of the animals restrict this theoretical number of states to a smaller number of *admissible states*.

The problem may be restated in the following way:

1. Find all admissible "double" states (subset of $\mathcal{S}$)

2. Construct 12 logical functions $c_1, c_2, \ldots c_6, m_1, m_2, \ldots m_6$ defined for each of the admissible states so that the constraints are fulfilled (synthesis of the mapping λ).

We may start by skipping some states that can be directly seen as not admissible (i.e. *forbidden*) due to the constraint 1 (*non-presence in the same room*) $\{(MRi, CRi), i = 0, 1, 2, 3, 4\}$ and as the door c_7 between 1 and 3 is always open, we must add $(MR1, CR3)$ and $(MR3, CR1)$. Constraint 2 (*free return home*) excludes all "double" states that contain as one component "single" states $MR2$ and $CR4$. In each of these cases, home of one animal is occupied by its rival and therefore, free return cannot take place. Thus the number of potential *admissible* states is reduced to $25 - 5 - 2 \cdot (5 - 1) - 2 = 10$. We shall start to deduce values of door controlling functions from the set of requirements that will prevent the animals from entering the forbidden states. (Only one animal is supposed to move). We can express the fact that when the mouse is in room 0, cat must be in the next step prohibited to enter $R0$, too, by condition $MR0 \cdot NCR0 \equiv 0$. Similarly for all forbidden states:

$$
\begin{array}{ll}
MR0 \cdot NCR0 \equiv 0 & NMR0 \cdot CR0 \equiv 0 \\
MR1 \cdot NCR1 \equiv 0 & NMR1 \cdot CR1 \equiv 0 \\
MR2 \cdot NCR2 \equiv 0 & NMR2 \cdot CR2 \equiv 0 \\
MR3 \cdot NCR3 \equiv 0 & NMR3 \cdot CR3 \equiv 0 \\
MR4 \cdot NCR4 \equiv 0 & NMR4 \cdot CR4 \equiv 0 \\
MR1 \cdot NCR3 \equiv 0 & NMR1 \cdot CR3 \equiv 0 \\
MR3 \cdot NCR1 \equiv 0 & NMR3 \cdot CR1 \equiv 0
\end{array}
\tag{3}
$$

two equations with $CR4$ and $MR2$ can be skipped as violating "free return". We are not interested in conditions for values of functions $c_1 \ldots m_6$ in forbidden states.

The additional supposition that only one element can move through the door at each moment is due to the fact that the maze is very small. E.g. if the cat goes through c_3 to $R0$ and the mouse through m_5 to $R3$, then the constraint 2 cannot be fulfilled. This supposition might be dropped for a larger maze with sufficiently distant home positions. In such a case, it would be necessary to add to above mentioned prohibiting conditions also the equations of the type $NMRi \cdot NCRi \equiv 0$.

To satisfy first condition $MR0 \cdot NCR0 \equiv 0$, let us calculate the product

$$MR0 \cdot NCR0 = \$c_3 \cdot CR2 \cdot MR0 + \$c_6 \cdot CR4 \cdot MR0 + \overline{\$c_4} \cdot \overline{\$c_1} \cdot CR0 \cdot MR0$$

Each of the conjuncts must be equal to logical zero. To annihilate first conjunct and supposing $(CR2, MR0)$ to be admissible state then $\$c_3$ must be 0 (or $\overline{\$c_3} = 1$). This can be guaranteed by closing the door c_3 for the state $CR2 \cdot MR0$. Then, to ease up further processing we may transcribe the first condition

$$MR0 \cdot NCR0 \equiv 0 : \qquad c_3 \cdot CR2 \cdot MR0$$

Second term was skipped because of $CR4$. Third term was skipped as $CR0 \cdot MR0 \equiv 0$. (i.e. $(CR0, MR0)$ is a forbidden state). Then, all relevant equations from (3) may be transcribed in the aforementioned way.

$$
\begin{array}{ll}
MR0 \cdot NCR0 \equiv 0 : & c_3 \cdot CR2 \cdot MR0 \\
MR1 \cdot NCR1 \equiv 0 : & c_1 \cdot MR1 \cdot CR0 \\
MR3 \cdot NCR3 \equiv 0 : & c_4 \cdot MR3 \cdot CR0 \\
MR4 \cdot NCR4 \equiv 0 : & c_5 \cdot MR4 \cdot CR3 \\
MR1 \cdot NCR3 \equiv 0 : & \text{nothing(because of } MR2) \\
MR3 \cdot NCR1 \equiv 0 : & c_1 \cdot CR0 \cdot MR3 \\
NMR0 \cdot CR0 \equiv 0 : & m_3 \cdot MR1 \cdot CR0 + m_6 \cdot MR3 \cdot CR0 \\
NMR1 \cdot CR1 \equiv 0 : & \text{nothing(because of } MR2) \\
NMR2 \cdot CR2 \equiv 0 : & m_1 \cdot MR0 \cdot CR2 \\
NMR3 \cdot CR3 \equiv 0 : & m_5 \cdot MR4 \cdot CR3 \\
NMR3 \cdot CR1 \equiv 0 : & m_5 \cdot MR4 \cdot CR1
\end{array}
\tag{4}
$$

Constraint 2 (*free return home*) will be taken into consideration in the following way. The mouse can return home (Room 4) when it is already at home (i.e. state $MR4$) or when it is in the states (rooms) $N1MR4$ *from* which it is possible enter $R4$ by mere passing one door or when $R4$ can be reached by passing 2 doors. Corresponding states are denoted $N2MR4$, then. For a larger maze, it would be necessary consider further potential "starting" states $N3MR4$ etc., but in the present maze candidates for free return states of the mouse may be described symbolically by $\Phi 1 = MR4 + N1MR4 + N2MR4$ and for the cat by $\Phi 2 = CR2 + N1CR2 + N2CR2$. Each of the terms (e.g. $N1MR4$ or $N2CR2$) will considered as a set of chains of symbols starting by some state of the animal proceeded by the symbol for an open door, next state and so on till the home state finishes the chain. (e.g. $CR0 \cdot \overline{c_1} \cdot CR1 \cdot \overline{c_2} \cdot CR2$).

Each system of home returning alternatives can be obtained by using the next state equations $(1) + (2)$ so that

a) conjuncts containing negation $\overline{\$ c_i}$ (of passing signal $\$ c_i$) are skipped (it expresses me re staying in the same room as before)

b) assertion of passing signals (e.g. $\$ c_1$) will be replaced by negation $\overline{c_1}$ of the door closing function c_1. It is clear that to enable the animal to pass the door c_1 (what will produce the signal $\$ c_1$), the door must be open i.e. $\overline{c_1} = 1$.

$N1MR4$ and $N1CR2$ will be constructed by direct applying the rules   **a)** and **b)** to functions $NMR4$ and $NCR2$ from (1) and (2) and $N2MR4$ and $N2CR2$ will be obtained by substituting states MRk and CRl occurring in $N1MR4$ and $N1CR2$ by right sides of $NMRk$ and $NCRl$ equations and again by applying rules **a)** and **b)**.

Using the procedure we find out that expressions $\Phi 1$ and $\Phi 2$ are equivalent to

$$
\begin{aligned}
\Phi_1' &= MR4 + MR0 \cdot \overline{m_4} \cdot MR4 + MR3 \cdot \overline{m_6} \cdot MR0 \cdot \overline{m_4} \cdot MR4 + \\
&\quad + MR1 \cdot \overline{m_3} \cdot MR0 \cdot \overline{m_4} MR4 \\
\Phi_2' &= CR2 + CR1 \cdot \overline{c_2} \cdot CR2 + CR0 \cdot \overline{c_1} \cdot CR_1 \cdot \overline{c_2} CR2 \\
&\quad + CR3 \cdot \overline{c_7} \cdot CR1 \cdot \overline{C_2} \cdot CR2
\end{aligned} \tag{5}
$$

Each of the terms in Φ_1' and Φ_2' represents a home returning alternative for each animal moving in the maze independently of the constraint 1. As the animals should have the possibility to return home even if they are both present in the maze , only some of alternatives are compatible.

Namely, such *chains are compatible* that do not cross (i.e. do not share the same room). Formally, we can cross-multiply terms of Φ_1' and Φ_2' and if we obtain a product corresponding to a forbidden state (we assume commutativity of symbols in the conjuncts), the chains corresponding to terms are not compatible. It is clear that shorter chains are more likely to survive this compatibility test and longer chains contain often previously defined chains (free return alternatives) and this should help in efficient elimination of incompatible chains. E.g. third terms in Φ_1' and Φ_2' are not compatible because of $MR0 \cdot CR0$ being a forbidden state. After having performed the formal multiplication, we find that there are 5 compatible pairs of chains:

$$P_1 = \{MR4, CR2\}$$
$$P_2 = \{MR4, CR1 \cdot \overline{c_2} \cdot CR2\}$$
$$P_3 = \{MR0 \cdot \overline{m_4} \cdot MR4, CR2\} \tag{6}$$
$$P_4 = \{MR0 \cdot \overline{m_4} \cdot MR4, CR1 \cdot \overline{c_2} \cdot CR2\}$$
$$P_5 = \{MR3 \cdot \overline{m_6} \cdot MR0 \cdot \overline{m_4} \cdot MR4, CR2\}$$

Admissible "double" states can be obtained as all cross products of "single" states of both animals for all pairs of compatible chains. According to this procedure we get following 7 admissible states:

$$(MR4, CR2), (MR4, CR1), (MR4, CR0), (MR4, CR3), (MR0, CR2), \tag{7}$$
$$(MR0, CR1), (MR3, CR2).$$

In fact, a more detailed analysis (similar to the processing of Φ_1 and Φ_2 but describing the states *into* which the system may develop starting from $MR4 \cdot CR2$) shows that though the state $(MR0, CR1)$ is admissible (i.e. fulfills constraint 2), it is not reachable (i.e. the controller fulfilling constraints 1 and 2 will never allow the animals to reach this state). Therefore, we might stop considering the conditions for this state, but, as the analysis was not displayed, we will not do that.

The pairs (6) of the compatible chains can be used to find states where the door closing functions achieve the value logical zero. If the admissible state was generated by some pair of chains then for this state all doors that are mentioned in the chains must be open. E.g. there is no requirement on value of any of control functions in the state $(MR4, CR2)$ generated by $P1$. On the other hand, there are requirements for both c_2 and m_4 to be opened $(\overline{c_2}, \overline{m_4})$ in the state $(MR0, CR1)$ generated by $P4$. Then, we can write directly the output functions for individual doors according to the equations (4) and (6).

$$c_1 = MR1 \cdot CR0 + CR0 \cdot MR3$$
$$\overline{c_2} = MR4 \cdot CR1 + MR0 \cdot CR1$$
$$c_3 = CR2 \cdot MR0$$
$$c_4 = MR3 \cdot CR0 + CR0 \cdot MR1$$
$$c_5 = MR4 \cdot CR3$$
$$m_1 = MR0 \cdot CR2 \tag{8}$$
$$m_3 = MR1 \cdot CR0$$
$$\overline{m_4} = MR0 \cdot CR2 + MR0 \cdot CR1$$
$$m_5 = MR4 \cdot CR3 + MR4 \cdot CR1$$
$$m_6 = MR3 \cdot CR0$$
$$\overline{m_6} = MR3 \cdot CR2$$

The admissible states are given by (7). The door closing functions are defined by (8) i.e. zeroes of c_i are required for states given by $\overline{c_i}$, logical ones for states given by c_i and, in unmentioned states, logical zeroes should be used as well as for unmentioned functions (e.g. c_6). System behaves dynamically according to (1) + (2).

Supposing that the animals are autonomous subjects, it is unrealistic to expect that maximally one of them may try to enter another room at each moment. But this supposition was used in the analysis made above. Therefore,its fulfilling must be forced out by additional closing of the doors whenever there exists a sign (e.g. $\#m_5$) that both animals are attempting to enter some doors at the same time. To treat simultaneous move of both animals, it is sufficient to add, to all output functions in assertive mode, the term $(\#c_1 + \#c_2 + \ldots \#c_7)(\#m_1 + \#m_2 + \ldots \#m_6)$ that blocks all the doors then.

CONCLUSION

The cat-and-mouse problem can be, in a natural way, solved using the context of automata and of logical functions. This approach can be applied also to a more complex example like the one introduced by G.J. Hoffmann and H. Wong-Toi (larger maze, in more dimensions; hunter-dog, dog-cat, mouse-cat) where there are more "animals" but antagonism is expressed only for the mentioned pairs. Then equations of type (3) will be written for antagonistic pairs only.

Replying to P.Caines about resistance of the approach to errors, we may suppose that errors can occur either due to the malfunction of sensors e.g. $\$m_5$ or by the fact that the animals are not in prescribed starting rooms or that there is more of them. In any of such cases, it is possible, by introducing additional output functions, at least to detect that something is wrong.

REFERENCES

[1] W. M. Wonham and P. J. Ramadge, On the supremal controllable sublanguage of a given language, SIAM J. Control Optimization vol. 25(1987), pp 637-659

Symbolic Supervisor Synthesis for the Animal Maze

Gérard J. Hoffmann[*]   Howard Wong-Toi[†]

Abstract. Ramadge and Wonham [1] gave algorithms for finding supervisors in their *supervisory control* framework. Their automatic synthesis techniques are implemented here using *binary decision diagrams* [2]. This technique of symbolic representation capitalizes on loose coupling between controlled plant components. We synthesize a supervisor for an extended animal maze problem of over 40 million states.

1. INTRODUCTION

The synthesis technique illustrated in this paper was first reported in [3, 4], where it was applied to the synthesis of supervisors for a wafer-manufacturing plant. This paper first repeats much of the material presented there, and then applies the technique to the synthesis of a supervisor for an extended animal maze problem.

The supervisory control theory of Ramadge and Wonham [5, 1] provides a formal framework for analyzing discrete event logic systems. The theory supplies algorithms for the automatic synthesis of supervisory controllers from their specifications. However an explicit implementation of these algorithms is often not practical because the size of the discrete state-space renders traditional computational methods infeasible.

We are aware of only a few attempts to implement automatic synthesis tools for discrete-event controllers [6, 7, 8]. Related approaches include controller design based on *iterative verification* where designs are successively automatically tested and improved [9].

In this paper we describe a synthesis implementation, and demonstrate its practical efficiency by synthesizing a supervisor for an extended animal maze problem. The implementation is based on a data structure known as a binary decision diagram (BDD) [2]. It is a compact symbolic representation method that avoids explicit enumeration of the entire discrete state-space. Research in the field of formal finite-state verification has shown BDDs can dramatically extend the capability of traditional algorithms [10, 11].

[*]Supported in part by Texas Instruments under contract No. 7554900. e-mail: hoffmann@isl.Stanford.EDU. Information Systems Laboratory, Stanford University, CA 94305, USA

[†]Supported by the Department of the Navy, Office of the Chief of Naval Research under Grant N00014-91-J-1901. This publication does not necessarily reflect the position or the policy of the U.S. Government and no official endorsement of this work should be inferred. e-mail: howard@cs.Stanford.EDU. Department of Computer Science, Stanford University, CA 94305, USA

2. PRELIMINARIES

Let Σ be a finite alphabet of symbols. Let Σ^* denote the set of all finite sequences over Σ. For an element $s = s_0, s_1, \ldots, s_{n-1} \in \Sigma^*$, we say $len(s)$, the length of s, is n. ϵ denotes the empty string. A finite string $t \in \Sigma^*$ is a *prefix* of s if $len(t) \leq len(s)$ and $t_i = s_i$ for $0 \leq i < len(t)$. A *language* L over Σ is any subset of Σ^*. The set of all languages over Σ is denoted $\mathcal{L}$. $pr(L)$ denotes the set of prefixes of L. L is *prefix-closed* iff $L = pr(L)$. We denote by $del(D)(L)$ the language obtained by removing all occurrences of symbols in D. Its inverse is $del^{-1}(D)(L') = \sup\{ L \mid del(D)(L) = L'\}$.

A *finite automaton* A is a tuple $\langle \Sigma, Q, I, F, \delta \rangle$, where Σ is a finite alphabet of transition symbols, Q is a finite set of automaton states, I is a set of initial states, $F \subseteq Q$ a set of final states, and $\delta : \Sigma \times Q \mapsto 2^Q$ is a partial transition function. If q' is in $\delta(q, \sigma)$, then it is possible to move from state q to q' on input symbol σ. The automaton is *deterministic* if $\delta(q, \sigma)$ is a singleton for every q and σ. An *accepting run* of A on the string $w \in \Sigma^*$ is a sequence s of $len(w) + 1$ states such that s_0 is in I, s_{i+1} is in $\delta(s_i, w_i)$ for $0 \leq i < len(w)$, and $s_{len(w)}$ is in F. The language generated by A, denoted $L(A)$, is the set of all strings w with accepting runs. A language L is *regular* if there is some automaton A such that $L(A) = L$. We use $|A|$ to denote the size of A, *i.e.* the number of states in A.

3. SUPERVISORY CONTROL

3.1 Supervisory Control Problem

Ramadge and Wonham's supervisory control theory [5, 1] uses formal languages of linear traces, or strings, to model both the plant and its specification. Each trace represents a sequence of events in a possible execution. The event set Σ is partitioned into *controllable* events Σ_c and *uncontrollable* events Σ_u. Intuitively uncontrollable events are always enabled, while controllable events can be prevented from occurring at any time. The uncontrolled plant or generator P is modeled as a pair (L_0, L_m) of languages of finite traces over Σ. L_0 is a prefix-closed language representing all possible partial executions of P, while $L_m \subseteq L_0$, its *marked language*, is a set of successfully *completed* traces.

A supervisor for P controls P's execution by observing its events and disabling possible events from occurring next. Formally, a *control mask* γ is any subset of Σ that contains Σ_u. Applying the mask γ means that every event in γ is enabled. Let Γ denote the set of all control masks. Given a plant P, a supervisor S is a pair (f, L_s) where f is a function $f : L_0 \to \Gamma$ and L_s is a marking language.

The plant's *supervised prefix language* L_{0f}, is given as: (1) $\epsilon \in L_{0f}$; (2) $w.\sigma \in L_{0f}$ iff $w \in L_{0f}$, $\sigma \in f(w)$ and $w.\sigma \in L_0$. Its marked supervised language is $L_{msf} = L_{0f} \cap L_m \cap L_s$, *i.e.* the supervised strings that are marked by both the plant and the supervisor. The controlled process P_f is the pair (L_{0f}, L_{msf}). If $L_{0f} = pr(L_{msf})$, then f is a *non-blocking* supervisor for P. Intuitively f is non-blocking if any partial execution allowed by f can be extended to a marked execution.

Supervisory Control Problem: Find a nonblocking supervisor f such that $L_{msf} \subseteq E$, where E is the specification language for the closed-loop behavior.[1]

3.2   Controllability Fixpoint Operator

Ramadge and Wonham [5] introduced the notion of *controllability* to characterize the supervised sublanguages of the plant. A language K is *controllable* with respect to L_0 iff $pr(K).\Sigma_u \cap L_0 \subseteq pr(K)$. The *supremal controllable sublanguage* of L with respect to L_0 is defined as $sup\ C(L_0, L) = \bigcup \{K : K \subseteq L$ and K is controllable wrt. $L_0\}$. The supervisory control problem has a solution iff $sup\ C(L_0, L_m \cap E)$ is non-empty. Therefore, solving the supervisory control problem reduces to effectively computing the greatest fixpoint of Ω for $L := L_m \cap E$. Ramadeg and Wonham prove $sup\ C(L_0, L)$ is the greatest fixpoint of the operator $\Omega : \mathcal{L} \mapsto \mathcal{L}$ defined as

$$\Omega(K) \quad = \quad L \cap sup\{T : T \subseteq \Sigma^*, T = pr(T) \text{ and } T\Sigma_u \cap L_0 \subseteq pr(K)\}$$

$$= \quad \{t : t \in L \text{ and } pr(t).\Sigma_u \cap L_0 \subseteq pr(K)\}$$

This suggests the iteration scheme $K_0 = L$, $K_{i+1} = \Omega(K_i)$ for $i \geq 0$, that converges at $K = \lim_{i \to \infty} K_i$.

We assume the plant and specification languages are regular. Let A_E be a deterministic finite-state automaton for E. Suppose the plant $P = (L_0, L_m)$ is given by an automaton A_P for $L_m \subseteq L_0$, where the language L_0 is the language obtained by considering all states final. The complexity of computing the greatest fixpoint of Ω is $O(|A_E|^2 . |A_P|^2)$.

3.3   Modular Systems

The plant is usually composed from modular subsystems. Formally, each plant subsystem $i \in \{1, 2, \ldots n\}$ can be modeled by a pair of languages (L_o^i, L_m^i) as described above. The components have local event alphabets Σ^i ($1 \leq i \leq n$) which are not necessarily disjoint. The alphabet of the global plant is $\Sigma := \Sigma^1 \cup \Sigma^2 \ldots \cup \Sigma^n$. The global plant is given by (L_o, L_m), where

$$L_o := del^{-1}(\Sigma - \Sigma^1)(L_o^1) \cap \ldots \cap del^{-1}(\Sigma - \Sigma^n)(L_o^n)$$

and L_m is similarly defined.

It is easy to see that the number of states in the global plant's automaton increases exponentially with n, the number of modular components. This fact is crippling when it comes to computation for realistic systems. Combatting this *state-space explosion* is currently an area of intensive research in the formal verification of finite-state systems. One suggested solution has been to encode the automata as BDDs, see section 4.. In this paper, just as in [3], we demonstrate that this technique is useful for synthesis as well as verification.

---

[1]In this paper we do not consider a minimally required behavior, but only a maximally tolerable behavior.

3.4   Fixpoint Operator with Logic Predicates

If Q is the state set of an automaton, let $Q' = \{q' | q \in Q\}$ be a set of states representing successor states. If the alphabet of edge labels is Σ, a next-state relation for the transition function can be expressed as a set of tuples δ in $Q \times \Sigma \times Q'$. Any set of n-tuples $T \subseteq X_1 \times \cdots \times X_n$ can be thought of as a boolean function $f : X_1 \times \cdots \times X_n \mapsto \{0,1\}$ where $f(t) = 1$ iff $t \in T$. We assume the uncontrollable events are expressed as the boolean formula Σ_u, and A_P's and A_E's next-state relations and final-state predicates encoded as the boolean functions δ_P and F_P, and δ_E and F_E respectively.

The fixpoint characterization of $sup\ C(L_0, L_m \cap E)$ is now re-expressed in terms of the next-state relations of the automata, rather than as languages. Let F_{PE} be the predicate $F_P \wedge F_E$, and let δ_{PE} be $\delta_P \wedge \delta_E$, the next-state relation of the product automaton. δ_{PE} is a boolean function over the domain $W = Q_P \times Q_E \times \Sigma \times Q'_P \times Q'_E$. The next-state relation for the least restrictive controller is the greatest fixpoint of the operator $\Omega : 2^W \mapsto 2^W$, defined as:

$$\Omega(Z) =$$
$$\delta_{PE}(s_1, t_1, e, s_2, t_2) \wedge Z(s_1, t_1, e, s_2, t_2)$$
$$\wedge\ \forall s, t.[\ \exists s_3, t_3, e_3.[Z(s, t, e_3, s_3, t_3) \vee Z(s_3, t_3, e_3, s, t)]\ ] \Rightarrow$$
$$\underbrace{[\ [\forall e', s'.[\ \Sigma_u(e') \Rightarrow [\delta_P(s, e', s') \Rightarrow \exists t'. Z(s, t, e', s', t')]\ ]\ ]}_{(1)} \wedge$$
$$\underbrace{BR(F_{PE}, Z)(s, t)}_{(2)}\ ]$$

$BR(F_{PE}, Z)$ is the least fixpoint of the operator $\Gamma : 2^V \mapsto 2^V$, where $V = Q_P \times Q_E$, and Γ is defined for every function Y over V as

$$\Gamma(Y) = F_{PE}(s_1, t_1) \vee \exists e, s_2, t_2.[\ Z(s_1, t_1, e, s_2, t_2) \wedge Y(s_2, t_2)\ ]$$

Intuitively, Ω iteratively removes events that may lead to an unpermitted uncontrollable event (Condition 1) and Γ enforces backwards-reachability (Condition 2), thereby guaranteeing all prefixes can be extended to a complete execution.

Theorem 1 *The fixpoint iteration scheme, $Z_{i+1} := \Omega(Z_i)$ with $Z_0 = \delta_{PE}$ effectively computes a function representing the next-state relation of an automaton for the supremal controllable sublanguage* $sup\ C(L_0, L_m \cap E)$.   ■

4.   BINARY DECISION DIAGRAMS

A *binary decision diagram* (BDD) [2] is an efficient data structure for representing boolean functions. It is a special form of rooted directed acyclic graph (DAG). A DAG can be used to represent a decision tree for a boolean function. This encoding is often far smaller than an explicit truth table. The value of the function for a particular variable assignment can be read by traversing the tree starting from the root, at each node branching according to the value of the variable labeling that node. Figure 1 shows a DAG representing $f = (x_1 \vee x_2) \wedge (x_3 \vee x_4)$. The variable assignment $(x_1 = 0, x_2 = 1, x_3 = 1, x_4 = 1)$ leads to a node marked

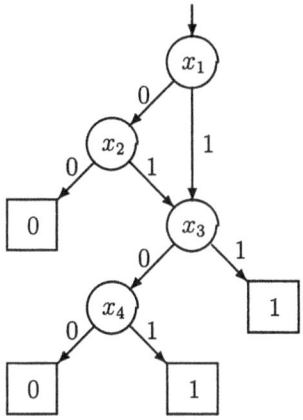

Figure 1: A DAG representing the boolean function $(x_1 \vee x_2) \wedge (x_3 \vee x_4)$.

1. Thus f is true under this variable assignment. Notice that the value of x_4 is irrelevant. The path followed symbolically represents the two variable assignments $(x_1 = 0, x_2 = 1, x_3 = 1, x_4 = 0)$ and $(x_1 = 0, x_2 = 1, x_3 = 1, x_4 = 1)$.

However a boolean function does not have a unique representation as a DAG. A BDD is a DAG satisfying the additional constraint that the occurrence of variables on every path from the root to a leaf obeys a given total order. The DAG in figure 1 is in fact a BDD with variable ordering $x_1 < x_2 < x_3 < x_4$. Bryant showed that for any total order on the variables, every boolean function is represented by a *unique* BDD respecting that order. He also gave efficient algorithms to perform standard boolean operations on BDDs. The complexity of finding the BDD for the logical AND, OR, or NOT of two BDDs is bounded by the product of the sizes of the two BDDs. To encode the fixpoints of the supervisory synthesis algorithms, we need the additional operation of *quantification*. The existential quantification formula $\exists x_i[f]$ can be read as "(f holds when x_i is FALSE) OR (f holds when x_i is TRUE)". We use Bryant's restriction algorithm for $f|_{x_i=0}$ and $f|_{x_i=1}$ to implement $\exists x_i[f]$ as $f|_{x_i=0} \vee f|_{x_i=1}$.

A BDD's size depends crucially on the chosen variable ordering. It is generally desirable for a variable ordering to bunch together variables that are highly interdependent. For example, suppose that $W_1 = \{x_{11}, \ldots, x_{1n_1}\}, \ldots, W_m = \{x_{m1}, \ldots, x_{mn_m}\}$ is a partition of the variables of f, and $f = f_1 \wedge \cdots \wedge f_m$ where each f_i depends only on variables in W_i. Let the size of the BDD for g be denoted $|g|$. Then $|f| = O(|f_1| + \cdots + |f_m|)$ under the variable ordering $x_{11} < \cdots < x_{1n_1} < \cdots < x_{m1} < \cdots < x_{mn_m}$.

5.   EFFICIENT IMPLEMENTATION

The boolean functions describing the plant and the specification are stored as BDDs in the implementation. As discussed in section 4., this representation is often logarithmically smaller than explicitly enumerating the relations. Controllers are synthesized as specified by the supervisory control problem by efficiently performing the iterations described in section 3.4.

The main synthesis program provides an interactive environment for entering low-level commands for creating and manipulating BDDs. It is built on top of the finite-state verification package *Ver*, designed and implemented by David Dill, Andreas Drexler, and Alan Hu. Their implementation uses Brace, Rudell, and Bryant's package for BDD manipulation [12].

6. APPLICATION: THE WORKSHOP EXERCISE

6.1 Animal Maze Description

The cat and mouse example was introduced by Ramadge and Wonham in [5], where more detailed formulation and explanation can be found. A cat and a mouse are placed in a maze (see figure 2), with the cat initially in room 2 and the mouse in room 4. Each doorway denoted by m_i and c_i, $1 \leq i \leq 7$ can be traversed exclusively by the mouse and by the cat, respectively, in the direction indicated. Each door (with the exception of c_7) can be opened or closed by means of control actions. A supervisor observes only discrete events generated by sensors in the doors indicating that an animal is just running through. In the original example the animals cannot leave the basic maze, *i.e.* the doorways subscripted by letters do not exist. The cat is initially in room 2 and the mouse in room 4.

The control problem is to find a feedback supervisor such that the closed-loop system satisfies the following specification:

1. The cat and the mouse never occupy the same room simultaneously.

2. It is always possible for the animals to return to their initial positions.

The problem solution should enable the animals to behave as freely as possible with respect to the given constraints.

For the purpose of the workshop we first extended the size of the animal maze. The basic maze floor plan is doubled into a north and a south wing. In addition it is replicated 8 times by adding 7 (similar) floors to the basic example. The animals can move north and south via c_n, m_n, c_s, m_s, as indicated in figure 2 as well as up and down via the stairways c_u, c_d, m_u, m_d. Initially the animals are on the first floor, in the south wing.

A second problem extension was obtained by adding two more animals to the maze, a rabbit and a dog. Specification 1 from above is now replaced by the following.

1. The cat and the mouse never occupy the same room simultaneously, nor do the cat and the dog, nor the dog and the rabbit.

There are doors for both the dog and rabbit precisely where there are doors for the cat. The rabbit starts in the north wing, in room 0, on the first floor and the dog in room 2 in the north wing, also on the first floor. The problem remains otherwise unchanged.

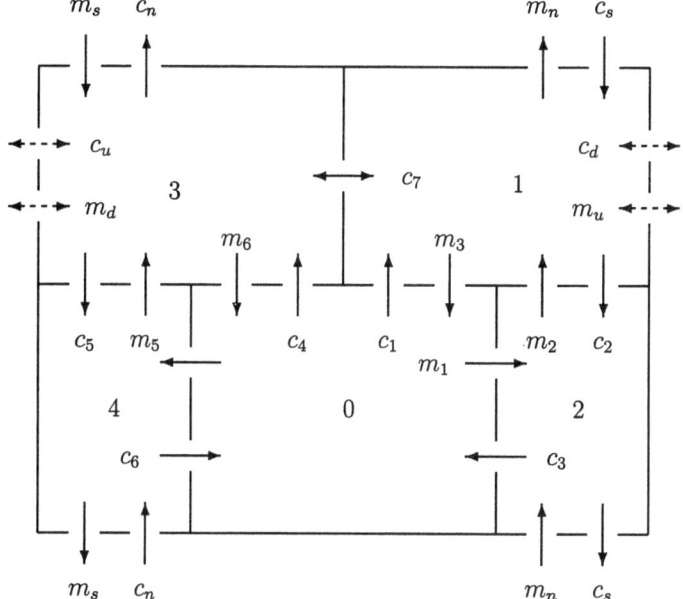

Figure 2: The maze

6.2 Modeling the Animal Maze

The animal maze is modeled as a collection of component processes. Each animal is modeled as a process with 5 states, one for each room. In addition a north-south wing-state for each animal monitors whether the animal is in the north or the south wing of the maze. The mouse process is different from the processes that model the other animals as the mouse uses different doors from the other animals.

The floor of each animal is monitored by a floor process for each animal. There are 8 floor-states for each animal. Table 1 reflects the described finite state model.

The specification is modeled in terms of a logical description of the forbidden states for the animals in the maze.

6.3 Sample Design Results

Supervisor synthesis results for the above problem description will be shown for three cases: the fairly simple original problem setup, the extended maze with 8 floors and two wings on each floor and the extended maze with a cat, a mouse, a dog and a rabbit. To illustrate the complexity of the described synthesis problem, we show in table 1 a state-space breakdown of the considered modules.

Clearly, a naive explicit computation method is bound to fail for the considered state-space size. Our implementation using BDD data structures yields the results shown in table 6.3. The computations were performed on a DecStation 5000/133. The BDD size refers to the number of nodes in the DAG representation

Maze Components	States
cat	2×5
mouse	2×5
rabbit	2×5
dog	2×5
cat-floor	8
mouse-floor	8
dog-floor	8
rabbit-floor	8
Max. Total	40.96×10^6

Table 1: State-space breakdown of the maze

of the synthesized controller. The table also lists the potential number of states. However not all states are reachable in the final solution. For example in the solution for the original maze only 6 states are reachable and in the extended maze with only a cat and a mouse there are 6285 out of 6400 states reachable. The result for the extended maze with 4 animals could be obtained by an appropriate variable ordering. The number of reachable states is 38.8×10^6 out of 40.96×10^6.

Case	CPU (s)	Mem.	BDD size	Pot. States	Reach. States
Original Maze	1.6	0.7	345	25	6
Extended Maze 1	13.8	1.3	1055	6400	6285
Extended Maze 2	1329	26.9	14778	40.96×10^6	38.8×10^6

Table 2: Results table

7. CONCLUSION

We outlined how supervisor synthesis could be performed symbolically by manipulating logic predicates represented as BDDs . The workshop example demonstrates the feasibility of supervisor synthesis for relatively large problems. An extension of the previous results to models and synthesis procedures that incorporate real-time is under investigation [13].

Acknowledgements

We thank Alan Hu and David Dill for explaining how their verifier works. We also thank Gene Franklin and David Dill for their generous support.

REFERENCES

[1] P.J. Ramadge and W.M. Wonham, The control of discrete event systems, *Proc. of the IEEE*, **77**(1):81–98, January 1989.

[2] R.E. Bryant, Graph-based algorithms for boolean function manipulation, *IEEE Transactions on Computers*, **C-35**(8):677–691, August 1986.

[3] G. Hoffmann and H. Wong-Toi, Symbolic synthesis of supervisory controllers, In *Proc. of 1992 American Control Conference*, pages 2789–2793, Chicago, IL, USA, June 1992.

[4] S. Balemi, G.J. Hoffmann, P. Gyugyi, H. Wong-Toi, and G.F. Franklin. Supervisory control of a rapid thermal multiprocessor. *Joint issue Automatica and IEEE Trans. Autom. Control on* Meeting the Challenge of Computer Science in the Industrial Applications of Control, 1993. to appear.

[5] W.M. Wonham and P.J. Ramadge, On the supremal controllable sublanguage of a given language, *SIAM J. Control Optim.*, **25**(3):637–659, May 1987.

[6] A. Benveniste P. Le Guernic, M. Le Borgne, Dynamical systems over Galois fields and DEDS control problems, In *Proc. of 30thConf. Decision and Control*, pages 1505–1509, Brighton, UK, December 1991.

[7] S.D. O'Young, TCT_talk: User's guide, Technical Report # 8915, Systems Control Group, Dept. of Electl. Engrg., Univ. of Toronto, Canada, October 1989.

[8] S. Lafortune and E. Chen, A relational algebraic approach to the representation and analysis of discrete event systems, In *Proc. of 1991 American Control Conference*, Boston, MA, USA, June 1991.

[9] A. Hsu, F. Eskafi, S. Sachs, and P. Varaiya, The design of platoon maneuver protocols for IVHS, Technical report, PATH Research Report, Institute of Transportation Studies, University of California at Berkeley, April 1991.

[10] J.R. Burch, E.M. Clarke, K.L. McMillan, D.L. Dill, and L.J. Hwang, Symbolic model checking: 10^{20} states and beyond, In *Proc. of 1990 IEEE Symposium on Logic in Computer Science*, Philadelphia, PA, 1990.

[11] E.M. Clarke and R.P. Kurshan, editors, *Computer-Aided Verification '90*, volume 3 of *DIMACS Series in Discrete Mathematics and Theoretical Computer Science*, American Mathematical Society, Association for Computing Machinery, 1991.

[12] Karl S. Brace, Richard L. Rudell, and Randal E. Bryant, Efficient implementation of a BDD package, In *Proc. of 27th ACM/IEEE Design Automation Conference*, pages 40–45, 1990.

[13] H. Wong-Toi and G. Hoffmann, The control of dense real-time discrete event systems, In *Proc. of 30th Conf. Decision and Control*, pages 1527–1528, Brighton, UK, December 1991.

The Cat-and-Mouse Problem with Least Delays

Petr Kozák*

1. INTRODUCTION

The formulation and solution of the cat-and-mouse problem within the formalism [1, 2, 3, 4, 5, 6, 7] are presented. True parallelism as well as the least delays between output discrete events are modelled. Both the plant and controller are strongly causal, therefore the results are relatively realistic with respect to a "real-world maze." The presented solution is minimally restrictive. The solution of the maze problem presented here is a sample example of utilization of the methodology [1, 2, 3, 4, 5, 6, 7], which can be used also for another classes of models and another control problems.

The output logical behaviour of the solution given here is equal to the supremal controllable sublanguage which is the solution of a corresponding logical discrete event system problem [8]. However in a general case, it is a subset of it. The reason of equality here is that the least delays of the maze model are non-zero.

As some concepts and notations are quite new and cannot be defined here due to the space limitation, we refer a reader to another paper in this volume [1], where an introduction and definitions can be found.

2. TIME MODEL OF THE CAT-AND-MOUSE SYSTEM

The models of the cat and mouse are given in the form of strong output laws μ_1 and μ_2, respectively (sections 5. and 7. in [1]). The model of the maze is the parallel interconnection of the animal models. It is modelled by a strong output law μ_o which is obtained from μ_1 and μ_2 using the set union operation. The models of cat and mouse are elementary discrete event systems (EDES) [5, 6, 7].

The EDESs are time models of discrete event systems modelling permissive control and the least delays between two subsequent output discrete events. The EDESs can be considered as a natural time extension of the logical discrete event systems modelled as formal languages in [8]. They are super ordinary and strongly causal [1].

*Czechoslovak Academy of Sciences, Institute of Information Theory and Automation, Pod vodárenskou věží 4, 182 08 Prague 8, Czechoslovakia; e-mail: kozak@cspgas11.bitnet
The work was partially supported by the grants ČSAV #27502 and ČSAV #27558, and Bell Canada research contract No. 3–254–188–10 (Univ. of Toronto). The paper was finished while the author was on leave at the Systems Control Group, Dept. of Electrical Engineering, University of Toronto, Toronto, Ontario, M5S 1A4 Canada; e-mail: kozak@odin.control.utoronto.ca

The cat-and-mouse system is modelled over the time set equal to non-negative real numbers $(\mathbb{R}_0^+, \leq)$ with the canonical order. The time moment zero corresponds to the "switching-on" the system. For an arbitrary set Σ, we denote by Σ^* the set of all finite sequences over Σ. The system model has several parameters:

- **Sets of output event labels.** $\Sigma_1 = \{c_1, c_2, \ldots, c_7\}$, $\Sigma_2 = \{m_1, m_2, \ldots, m_6\}$
- **Sets of labels of controllable events.** $\Sigma_{c,1} = \Sigma_1 \setminus \{c_7\}$, $\Sigma_{c,2} = \Sigma_2$, where $\setminus$ is the set difference.
- **Sets of input event labels.** $U = \{E_\sigma : \sigma \in \Sigma_{c,1} \cup \Sigma_{c,2}\} \cup \{D_\sigma : \sigma \in \Sigma_{c,1} \cup \Sigma_{c,2}\}$. The label E_σ models enabling and D_σ disabling of the events labelled by σ.
- **Output logical behaviours.** They are described by languages $L_1 \subseteq \Sigma_1^*$ and $L_2 \subseteq \Sigma_2^*$ which are equal to the languages generated by the automata $\mathcal{G}_c$ and $\mathcal{G}_m$ modelling the uncontrolled logical behaviour in [8]. They are prefix closed, i.e. $\mathrm{pref}(L_1) = L_1$ and $\mathrm{pref}(L_2) = L_2$.
- **Mappings describing least delays between two subsequent output discrete events.** $\xi_1 : \Sigma_1^* \times \Sigma_1 \mapsto \mathbb{R}_0^+$ and $\xi_2 : \Sigma_2^* \times \Sigma_2 \mapsto \mathbb{R}_0^+$. The value of $\xi_1(w, \sigma)$ is interpreted as the least delay between the last event (or the time moment 0 if there is no last event, i.e. in the case of $w = \epsilon$, where ϵ is the empty word) and an event labelled by σ, supposing that the logical output behaviour w has been observed previously. The least delay means here the infimum of all possible delays between the last event and a future event labelled by σ supposing that the logical observation w is observed.

The least delays (mappings ξ_1 and ξ_2) can be easily determined as ratios of the maximal possible speeds of animals and of distances between the doors. All delays are non-zero in this case. The least delays between 0 and the first output discrete event (i.e. $\xi_1(\epsilon, \cdot)$ and $\xi_2(\epsilon, \cdot)$) depend on the initial position of the animals within the rooms and on their initial vector of speed. If no information about it is available then they should be equal to zero.

Before we define the output laws modelling the maze we shall introduce two notations related to the least delays and the way of control. Let $\mathcal{M}$ be the set of all mappings in $\mathcal{F}(\mathbb{R}_0^+, 2^{U \cup \Sigma_1 \cup \Sigma_2})$ satisfying (4) in [1]. To simplify notation, an arbitrary sequence of singleton sets $\{a\}\{b\}\{c\}\ldots$ denotes from now on also the sequence of set elements $abc\ldots$

For all $f \in \mathrm{pref}(\mathcal{M})$ and $t \in \mathbb{R}_0^+$ we define

$$E_1(f, t) = \{\sigma \in \Sigma_1 : \xi(s(m_{\Sigma_1} \circ f), \sigma) < t - \tau\},$$

where

$$\tau = \max\{t' \in \mathrm{Dom}(f) : t' = 0 \vee f(t') \cap \Sigma_1 \neq \emptyset\}.$$

The set $E_1(f, t)$ is interpreted as the labels of cat output events that could be observed at the time t after observation of f taking into account only least delays, not e.g. control, prescribed output logical behaviour, etc. Analogously, we can introduce $E_2(f, t)$ for the mouse.

For all $\sigma \in \Sigma_1 \cup \Sigma_2$ and $f \in \mathrm{pref}(\mathcal{M})$ we define

$$\eta(\sigma, f) = \begin{cases} 1 & \text{if for all } t \in \text{Dom}(f), \ f(t) \cap \{E_\sigma, D_\sigma\} = \emptyset, \\ 0 & \text{if } D_\sigma \in f(\tau), \\ 1 & \text{otherwise,} \end{cases}$$

where $\tau = \max \{t \in \text{Dom}(f) : f(t) \cap \{E_\sigma, D_\sigma\} \neq \emptyset\}$.

The mapping $\eta(\sigma, f)$ determines for each event label σ whether it is enabled (value 1) or disabled (value 0) with respect to control after observation of f. We suppose that all events are enabled at the time of switching-on the system. Only the event labels in $\Sigma_{c,1} \cup \Sigma_{c,2}$ can be controlled.

Now we are ready to formulate the strong output law μ_1 modelling the cat. Let $\mathcal{M}_1$ be the set of all mappings in $\mathcal{F}(\mathbb{R}_0^+, 2^{U \cup \Sigma_1})$ satisfying (4) in [1]. We define for all $f \in \mathcal{M}_1$ and $t \in \mathbb{R}_0^+$

$$\mu_1(f, t) = \{\emptyset\} \cup$$
$$\{\{\sigma\} : \sigma \in \Sigma_1 \wedge s(m_{\Sigma_1} \circ f \mid \mathcal{J}_t^s)\sigma \in L_1 \wedge$$
$$\wedge \ \sigma \in E_1(f \mid \mathcal{J}_t^s, t) \wedge \eta(\sigma, f \mid \mathcal{J}_t^s) = 1 \}.$$

The behaviour of the cat is defined in the way described in [1] section 5. (5), i.e. as

$$\mathcal{G}_1 = \{f \in \mathcal{M}_1 : \forall t \in \mathbb{R}_0^+, \ m_{\Sigma_1} \circ f(t) \in \mu_1(f, t) \}. \tag{1}$$

There are four terms in the definition of μ_1. The first term defines that only the event labels in Σ_1 can be generated. The second one ensures that only the words in L_1 are generated. The third term is responsible for the delays between output events. The last term states that only the event labels enabled by means of control can be generated.

The model of the mouse can be defined in an analogous way using μ_2 and $\mathcal{M}_2 \subseteq \mathcal{F}(\mathbb{R}_0^+, 2^{U \cup \Sigma_2})$.

The output law μ_o describing the maze is defined as follows. For all $f \in \mathcal{M}$ and $t \in \mathbb{R}_0^+$

$$\mu_o(f, t) = \mu_1(m_A \circ f, t) \cup \mu_2(m_B \circ f, t),$$

where $A = U \cup \Sigma_1$ and $B = U \cup \Sigma_2$.

Let $Y = \Sigma_1 \cup \Sigma_2$. The input/output behaviour $\mathcal{G}$ of the maze is defined analogously

$$\mathcal{G} = \{f \in \mathcal{M} : \forall t \in \mathbb{R}_0^+, \ m_Y \circ f(t) \in \mu_o(f, t) \}.$$

It can be verified that for all $f \in \text{pref}(\mathcal{M})$ such that

$$\forall t \in \text{Dom}(f), \ m_Y \circ f(t) \in \mu_o(f, t)$$

it holds that $f \in \text{pref}(\mathcal{G})$.

We outline briefly the proof. There exists a mapping $g \in \mathcal{M}$ such that $g \mid \text{Dom}(f) = f$ and for all $t \in \mathbb{R}_0^+ \setminus \text{Dom}(f)$ it holds $m_Y \circ g(t) = \emptyset$. The definitions of μ_1, μ_2, and μ_o imply that $g \in \mathcal{G}$. Therefore $f \in \text{pref}(\mathcal{G})$.

Moreover, it can be verified that for all $u \in m_U \circ \mathcal{M}$ such that $m_U \circ f = u \mid \text{Dom}(f)$ we can find some g defined above that satisfies additionally $m_U \circ g = u$.

Therefore the sufficient condition (6) in [1] is satisfied and the model is well-formed.

Let us note that the logical output behaviour of $\mathcal{G}$ is

$$s(m_Y \circ \mathcal{G}) = \{w \in (2^Y \setminus \{\emptyset\})^* : s(m_{\Sigma_1} \circ w) \in L_1 \wedge s(m_{\Sigma_2} \circ w) \in L_2\}.$$

The elements of the sequences w are non-empty sets. Non-singleton sets describe simultaneous events, i.e. simultaneous jumps of the animals through the doors. Let us note that $m_{\Sigma_1} \circ w$ can have elements equal to the empty set. But they are removed by the operation "s".

3. CONTROL PROBLEM FORMULATION

The control problem is to find a feedback controller $\mathcal{S}_c$ for the given system $S = (\mathbb{R}_0^+, \leq, U, Y, \mathcal{G})$ such that the feedback interconnection of S and $\mathcal{S}_c$ satisfies certain constraints. The necessary and sufficient condition of existence of a feedback controller is existence of certain controllable behaviour [1, 4]. As the controller can be constructed on the basis of a controllable behaviour, the control problem can be reformulated equivalently as the problem of finding a controllable behaviour $\mathcal{Z}$ of the given system S which satisfies certain constraints [1, 4].

The specification of the desired behaviour of the maze can be formulated in the form of the control problem presented in section 6. [1]. We shall continue in the way given by the controller design scheme presented in the same section.

Let us denote by ξ_1 the transition mapping of $\mathcal{G}_c$ (describing the logical output behaviour of the cat in [8]) and by ξ_2 the transition mapping describing the mouse. The state spaces of automata equal to the names of rooms. The initial state of the cat is 2, and 4 for the mouse. For each word in $w \in L_1$ and $w \in L_2$ the values of $\xi_1(w, 2)$ and $\xi_2(w, 4)$ describe the rooms reached by the cat or mouse, respectively, after observation of w.

The parameters of the control problem are defined as follows. $L = s(m_Y \circ \mathcal{G})$ and

$$L_m = \{w \in L : \xi_1(s(m_{\Sigma_1} \circ w), 2) = 2 \wedge \xi_2(s(m_{\Sigma_2} \circ w), 4) = 4\}.$$

The marked language L_m consists of all words in L which correspond to the animal paths that move them again to their initial positions.

The legal behaviour $L_g \subseteq L_m$ is defined as follows

$$L_g = \{w \in L_m : \forall v \in \text{pref}(w), \ \xi_1(s(m_{\Sigma_1} \circ v), 2) \neq \xi_2(s(m_{\Sigma_2} \circ v), 4) \}.$$

The set of legal logical output observations consists of all observations in L_m in which the animals are in different rooms.

The minimal acceptable output logical behaviour L_a is the empty word. Nevertheless, we show minimally restrictive solution of the problem [1, 4], which has the logical output behaviour equal to the supremal controllable sublanguage solving the corresponding control problem in [8].

4. SOLUTION

It is proved that the output logical behaviour $s(m_Y \circ Z)$ of each solution Z is controllable language, supposing that a simultaneous event label (i.e. non-singleton set) is controllable if at least one label forming this simultaneous label is controllable. The output logical behaviour of Z must be also L_m-closed. These two properties show that the time extension is consistent with the original formulation [8]. The controller must take into account the true parallelism and the least delays, therefore the control strategy is quite different from the original one. However in certain cases, the output logical behaviour of Z can be equal to the supremal controllable sublanguage of the original problem. It is a subset in general. It depends basically on the least delays. It can be shown for the case of the maze, where the least delays are non-zero apart from the initial ones, that it can be equal to the supremal controllable sublanguage.

The control strategy can be written in the form of a strong control law μ_c : $\mathcal{G} \times \mathbb{R}_0^+ \mapsto 2^{2^U} \setminus \{\emptyset\}$. A general form of this control law is presented in [7]. If the system parameters are putted into this general form, the control law can be defined as follows. For all $f \in \mathcal{G}$ and $t \in \mathbb{R}_0^+$ the set of all possible control actions is

$$\mu_c(f,t) = \{x \in 2^U : (2) \wedge (3) \wedge (4) \wedge (5) \wedge (6) \}$$

where

$$(\text{cat\_room} \neq 2 \wedge m_5 \in \text{mouse\_del}) \Rightarrow \eta(m_5, u) = 0, \tag{2}$$
$$(\text{mouse\_room} \neq 4 \wedge c_3 \in \text{cat\_del}) \Rightarrow \eta(c_3, u) = 0, \tag{3}$$
$$(\text{cat\_room} = 3 \wedge c_5 \in \text{cat\_del}) \Rightarrow \eta(c_5, u) = 0, \tag{4}$$
$$(\text{mouse\_room} = 0 \wedge m_1 \in \text{mouse\_del}) \Rightarrow \eta(m_1, u) = 0, \tag{5}$$
$$(\text{cat\_room} = 2 \wedge \text{mouse\_room} = 4 \wedge c_3 \in \text{cat\_del} \wedge m_5 \in \text{mouse\_del}) \tag{6}$$
$$\Rightarrow (\eta(c_3, u) = 0 \vee \eta(m_5, u) = 0),$$

and

$$\begin{aligned}
\text{cat\_room} &= \xi_1(s(m_{\Sigma_1} \circ f \,|\, \mathcal{J}_t^s), 2), \\
\text{mouse\_room} &= \xi_2(s(m_{\Sigma_2} \circ f \,|\, \mathcal{J}_t^s), 4), \\
\text{cat\_del} &= \{\sigma \in \Sigma_1 : \forall t' \in \mathbb{R}_0^+, \ t < t' \Rightarrow \sigma \in E_1(f \,|\, \mathcal{J}_t^s, t') \}, \\
\text{mouse\_del} &= \{\sigma \in \Sigma_2 : \forall t' \in \mathbb{R}_0^+, \ t < t' \Rightarrow \sigma \in E_2(f \,|\, \mathcal{J}_t^s, t') \},
\end{aligned}$$

and $u \in m_U \circ \mathcal{G} \,|\, \mathcal{J}_t$ such that

$$u \,|\, \mathcal{J}_t^s = m_U \circ f \,|\, \mathcal{J}_t^s \wedge u(t) = x.$$

The symbol "cat_room" denotes the room reached by the cat after the observation $f \,|\, \mathcal{J}_t^s$. The symbol "cat_del" denotes all event labels in Σ_1 which could occur at any time moment greater than t after observation of $f \,|\, \mathcal{J}_t^s$ considering the least delays only. Analogously for "mouse_room", "mouse_del", and the mouse.

The right hand side of the implications are interpreted as the requirements for control, i.e. they determine what events must be disabled after the interval $\mathcal{J}_t$.

E.g. (6) can be rewritten as the following rule:

If the room occupied by the cat is 2 and the room occupied by the mouse is 4 and both c_3 and m_5 can occur (with respect to the least delays) at any time moment greater than t then at least one of the event labels c_3 and m_5 must be disabled after $\mathcal{J}_t$.

It means that it is not necessary to wait with the disablement till some moment t when the left hand side of the implication is true. But the events can be disabled earlier. The rules forming the control law can be rewritten in a straightforward manner into a computer program.

The set $\mathcal{Z}$ (corresponding to the behaviour of closed-loop system) is defined as

$$\mathcal{Z} = \{ f \in \mathcal{G} : \ \forall t \in \mathbb{R}_0^+, \ m_U \circ f(t) \in \mu_c(f, t) \ \}.$$

It is shown [7] that the set $\mathcal{Z}$ satisfies the condition (8) in [1]. Therefore $\mathcal{Z}$ is a strongly controllable behaviour of $\mathcal{S}$. It also satisfies the constraints of the control problem [7] and therefore it is a solution of the problem.

It can be shown that for all $f \in \mathcal{G} \setminus \mathcal{Z}$ there does not exist any solution $\mathcal{Z}'$ of the given control problem such that $f \in \mathcal{Z}'$ and $\mathcal{Z} \subset \mathcal{Z}'$. Solutions having this property are called minimally restrictive, as they describe all possible ways in which the given system can be controlled such that the given constraints are satisfied.

5. CONCLUSION

The non-zero least delays simplify the formulation of control problem and the structure of the controller. However, arbitrary least delays are considered in the model [7].

The presented results proved within the framework [1, 2, 3, 4, 5, 6, 7] form a first step in development of a qualitative discrete event systems control theory suitable for applications in real-world problems. A next step is introducing communication delays (some preliminary results are in [6]). Future research will be aimed to hierarchical and modular techniques and to the modelling of more complicated rules of discrete event generation. Also possibilities of application of the framework to more general classes of systems will be addressed.

REFERENCES

[1] P. Kozák. A unifying framework for discrete event system control theory. *In this volume*, pages 95–110.

[2] P. Kozák. Discrete events and general systems theory. *Int. J. Systems Science*, 23(9):1403–1422, September 1992.

[3] P. Kozák. On feedback controllers. *Int. J. Systems Science*, 23(9):1423–1431, September 1992.

[4] P. Kozák. On controllable behaviours of time systems. Technical Report # 1739, Czechoslovak Academy of Sciences, Institute of Information Theory and Automation, Prague, Czechoslovakia, 1991. submitted.

[5] P. Kozák. Supervisory control of discrete event processes: A real-time extension. Technical Report # 1692, Czechoslovak Academy of Sciences, Institute of Information Theory and Automation, Prague, Czechoslovakia, 1991, revised 1992. submitted.

[6] P. Kozák. Control of elementary discrete event systems: Synthesis of controller with non-zero decision time. In D. Franke and F. Kraus, editors, *Proc. of the 1st IFAC Symposium on Design Methods of Control Systems*, pages 457–462, Zurich, Switzerland, September 1991. Pergamon Press, Oxford.

[7] P. Kozák. Supervisory control of parallelly interconnected elementary discrete event processes. *submitted*, 1992.

[8] W.M. Wonham and P.J. Ramadge. On the supremal controllable sublanguage of a given language. *SIAM J. Control Optim.*, 25(3):637–659, May 1987.

SUPERVISORY CONTROL WITH
VARIABLE LOOKAHEAD POLICIES:
ILLUSTRATIVE EXAMPLE

Sheng-Luen Chung*   Stéphane Lafortune *   Feng Lin [†]

Abstract.
We use the Limited Lookahead Policy (LLP) supervisory control scheme of [2, 3, 4, 1] to solve the Cat-and-Mouse problem of [5]. The solution employs the Variable Lookahead Policy (VLP) algorithms of [4]. These algorithms provide an efficient implementation technique of the LLP scheme.

1. INTRODUCTION

Consider a discrete event system (or DES), denoted by G and with corresponding closed language $L(G)$ and marked language $L_m(G)$, that is being controlled by dynamically disabling/enabling events after the execution of each event by the controlled system. G is being controlled in order to satisfy specifications for marked traces represented by the "legal" language $K \subseteq L_m(G)$. In supervisory control with limited lookahead policies (or LLPs), as depicted in figure 1, the control action $\gamma_{l,a}^N(s)$ after a given trace of events s has been executed is calculated on-line on the basis of an N-step ahead projection of the behavior of the DES under consideration; this procedure is repeated after the system executes any one of the enabled events. The resulting closed-loop behavior is denoted $L(G, \gamma_{l,a}^N)$. This is in contrast with the "conventional" supervisory control paradigm (cf. [5, 6]) where the complete control policy is calculated off-line using automaton models of the DES and of the legal behavior. As discussed in [2], LLPs allow us to control time-varying DES's and they also provide a means for dealing with the computational complexity of supervisor synthesis for DES's with large state spaces.

Reference [2] presents a detailed study of the LLP scheme along with its optimality properties (in particular, in terms of N, the size of the "rolling window"). References [3, 4]are specifically concerned with the computation of the LLP controls. To compute these controls, one must calculate after the execution of each event by the system the supremal controllable sublanguage of a finite language with respect to another finite larger language. The latter language, denoted by $L(P)$, is the set of all traces of events that the system can generate in the next

*Department of Electrical Engineering and Computer Science, University of Michigan, Ann Arbor, MI 48109-2122, USA. Research supported in part by the National Science Foundation under grant ECS-9057967, with additional support from DEC and GE.

[†]Department of Electrical and Computer Engineering, Wayne State University, Detroit, MI 48202, USA. Research supported in part by the National Science Foundation under grant ECS-9008947.

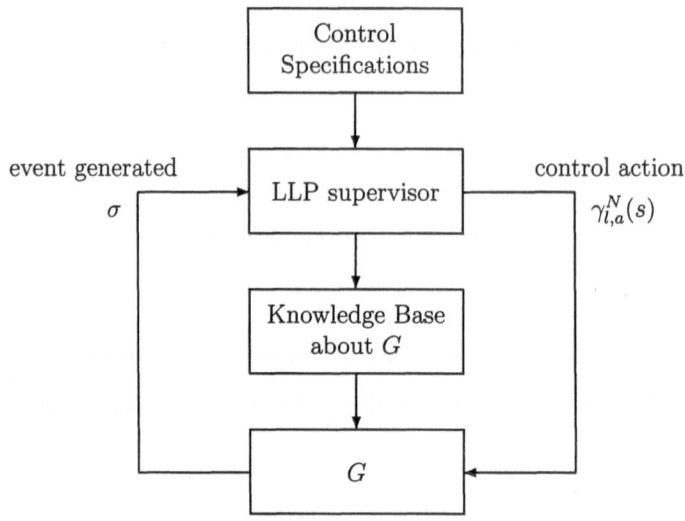

Figure 1: Limited Lookahead Supervisory Control

N steps, while the former language, denoted by L, is the subset of these traces that are legal according to the specifications on event ordering.

In [3], the required supremal controllable sublanguage computation is approached as an optimal control problem with a $0/\infty$ cost structure. A special control cost assignment scheme is adopted to reflect the following requirements: (1) the supervisor will not disable an uncontrollable event; and (2) the behavior of the supervised system is non-blocking. This optimal control problem is solved by using dynamic programming over a state space X consisting of the nodes of a tree representation of the aforementioned finite language $L(P)$. Each state x in X is classified as a member of any of the following three groups: $X_{illegal}$, X_m, $X_{transient}$, depending on the legality classification of the associated node as: illegal, legal and marked, or legal and unmarked, respectively. The subset of states in X_m that have no uncontrollable continuations in $L(P)$ is denoted by X_{mc}. The optimal cost-to-go function of this optimal control problem is defined by $V_{l,a}^N : X \longrightarrow \{0, \infty\}$, where the subscript a stands for the "attitude" adopted regarding the legality of the traces of length N in X (see below). According to Theorem 5 of [3], the control action $\gamma_{l,a}^N(s)$ equals the set of all next events $\sigma \in \Sigma_{L(G)}(s)$ (the active set of $L(G)$ at s) whose $V_{l,a}^N(\sigma) = 0$.

Section 4.2 of [3] presents a backward dynamic programming algorithm for the calculation of the function $V_{l,a}^N$ over a given N-level tree. This algorithm emphasizes the recursiveness between such calculations from N-level tree to N-level tree as the N-step window rolls with the execution of one (enabled) event by the DES, in addition to the inherent recursiveness from level to level of the tree at a given step. Reference [4] presents an improved algorithm that is based upon a forward calculation procedure over the state space (i.e., N-level tree) of interest.

This forward procedure may avoid the explicit consideration of all the nodes (or states) of the N-level tree, while still permitting step-to-step (i.e., tree-to-tree) recursiveness. The forward search ends whenever a control decision can be made unambiguously with respect to the future behavior or whenever the boundary of the N-level tree is reached, whichever comes first.

A *Variable Lookahead Policy* (or VLP) is a limited lookahead supervisory control policy whose on-line implementation at each step is by means of a forward search technique over the N-level tree of interest. Thus VLPs are an efficient implementation technique of the LLP scheme. In particular, VLPs share all the properties of the LLP scheme that are presented in [2]. We call a *VLP algorithm* the algorithm employed by a VLP for the required forward search over the N-level tree.

We have proposed in [4] four VLP algorithms termed and denoted: no-bound-V_v, undecided-V_{vu}^N, conservative-$V_{v,cons}^N$, and optimistic-$V_{v,optm}^N$ (cf. [4, 1]). These VLP algorithms are all based on the above-described forward search technique for computing the cost-to-go function in a recursive manner. The differences among them are: (i) the limitation on the maximal allowed search depth (window size) N, and (ii) the attitude adopted to resolve the system uncertainty beyond N steps, when the search hits the boundary. (The conservative and optimistic attitudes are described in [2]; the undecided attitude is described in [4].)

In this paper, we concentrate on the solution of the Cat-and-Mouse problem of [5] by the VLP algorithm with undecided attitude, V_{vu}^N, and then comment on the other variations and make suggestions for potential extensions. The Cat-and-Mouse problem is used here for illustrative and comparison purposes. In its original form, it does not have any of the attributes that would favor an on-line limited lookahead solution over the conventional off-line solution, such as large state space, recognizer of K difficult to build, or G (and/or K) time-varying. The reader is referred to [4] for an example involving a substantially larger number of states. Before proceeding to the specifics of the Cat-and-Mouse example, we summarize some results from [4] regarding the VLP algorithms to be employed.

2. THE VLP ALGORITHMS

We first state the VLP algorithm where the boundary condition set by (1) below corresponds to the "undecided" attitude, whose cost is denoted by U, where $0 \leq U \leq \infty$. Thus $V_{vu}^N : X \longrightarrow \{0, U, \infty\}$.

VLP Algorithm V_{vu}^N:
(define function cost-to-go (x) ; this function returns $V_{vu}^N(x)$.
case:

1. $|x| = N$:

$$V_{vu}^N(x) = \begin{cases} \infty & x \in X_{illegal} \\ U & \text{otherwise.} \end{cases} \tag{1}$$

2. $|x| < N \wedge x \in X_{mc}$: $V_{vu}^N(x) = 0$; return.

3. $|x| < N \wedge x \in X_{illegal}$: $V_{vu}^N(x) = \infty$; return.

4. $|x| < N \wedge x \notin X_{mc} \wedge x \notin X_{illegal}$:
case:

(a) $\Sigma_{L(P)}(x) \cap \Sigma_u \neq \emptyset$:
let $V_{vu}^N(x) = 0$
(do for all $(\sigma_u \in \Sigma_{L(P)}(x) \cap \Sigma_u)$ until $V_{vu}^N(x\sigma_u) = \infty$:

(cost-to-go $(x\sigma_u)$)
$V_{vu}^N(x) = max(V_{vu}^N(x\sigma_u), V_{vu}^N(x)))$

return.

(b) $\Sigma_{L(P)}(x) \cap \Sigma_u = \emptyset$:
let $V_{vu}^N(x) = \infty$
(do for all $(\sigma_c \in \Sigma_{L(P)}(x) \cap \Sigma_c)$ until $V_{vu}^N(x\sigma_c) = 0$:

(cost-to-go $(x\sigma_c)$)
$V_{vu}^N(x) = min(V_{vu}^N(x\sigma_c), V_{vu}^N(x)))$

return.)   ∎

The VLP algorithm with the conservative (optimistic, respectively) attitude is exactly the same as the above except that the U in (1) is replaced by ∞ (0, respectively). These two versions of the VLP algorithm are denoted by $V_{v,a}^N$, where a stands for *cons* or *optm*. However, V_{vu}^N has the advantage of being able to indicate to what extent the final control action is affected by the uncertainty of system behavior beyond N steps.

We now recall two important results from [4]. The first result is related to the correctness of the above algorithm.

Theorem 1 *The cost-to-go $V_{v,a}^N$ derived by either the conservative or the optimistic attitude can be obtained from that of $V_{vu}^N(x)$ derived by the undecided attitude, through a simple substitution. In addition, under the same attitude, for all x in the N-step tree, $V_{v,a}^N(x)$ is equal to $V_{l,a}^N(x)$, the cost-to-go derived by the backward dynamic programming technique of [3].*   □

As suggested by (1), when the forward search hits the boundary, an uncertain cost-to-go will be rippled back to the upstream nodes. This implies that: (i) the final decision on the control action can be affected by the adopted attitude, and (ii) the derived path up to the bouncing point may make no decisive contribution to the control action; derivation on other paths (thus extra computation) is likely to follow. In general, a larger window size N makes it less likely to hit the boundary, while risking increased computational complexity. Such a dilemma is mitigated by the following result. We define K_{mc} to be the set of traces in K whose active set does not contain any uncontrollable event.

Theorem 2 *Under the assumption that $K^\uparrow \neq \emptyset$, if*

$$N' := max\{|t| : (\exists s \in \overline{K})(st \in K_{mc} \cup (L(G) - \overline{K}) \wedge (\forall p < t)sp \notin K_{mc} \cup (L(G) - \overline{K})\}$$

exists and is finite, then the VLP algorithm with $N > N'$ will never hit the boundary at the Nth layer. The control action thus derived is independent of *the adopted attitude. Furthermore, the on-line solution is the same as the optimal off-line solution, namely $\overline{K^\uparrow}$, the prefix-closure of the supremal controllable sublanguage of K.*   □

Effectively, when the sufficient condition is satisfied, the boundary condition as specified by (1) can be removed. The resultant "no-bound" algorithm, denoted V_v, expands traces *only* as necessary. Without *a priori* information on the system behavior, this no-bound algorithm is computationally optimal.

The proposed "VLP implementation" of the LLP scheme is computationally efficient in calculating the on-line control actions due to the following two factors: (i) the forward search procedure employed at each step, and (ii) the structural similarity between the N-step projections at successive steps. Regarding (i), the VLP algorithms in general only expand the most relevant nodes during the forward search procedure, while ignoring those nodes that make no contribution to the final decision for the control action. To be more precise, nodes reached by the following types of events can be ignored: (a) events that are siblings (in the current active set) of an uncontrollable event leading to a node whose cost-to-go is either U or ∞; and (b) events that are controllable siblings of a controllable event leading to a node whose cost-to-go is 0, when there is no uncontrollable event present in the current active set. Moreover, nodes that are successors of nodes in $X_{illegal}$ and X_{mc} can also be ignored.

Regarding (ii) above, the structural similarity between successive steps suggests a strong correlation in the process of deciding successive control actions. For those nodes generated in previous steps and whose costs-to-go are either 0 or ∞, the cost-to-go should not change as the system moves from one step to another. This "final value property" allows re-utilization of calculations between successive steps. It is formalized in Theorem 4.2 of [4]. Observe that this property is meant to be used for the nodes of successive N-level trees, not for the true states of the system. In the general formulation of the LLP scheme (cf. [2]) considered here, the true system state is assumed to be unknown by the supervisor; the only knowledge available to the supervisor is the aforementioned language $L(P)$ and the legality properties of the traces in this language. (This is why $L(P)$ is represented by a tree generator.)

3.   SOLUTION OF THE CAT-AND-MOUSE PROBLEM

When one applies the VLP algorithms, nodes in the N-level tree under consideration are not all generated *a priori* but only when needed in the forward search procedure. When generated, each node has to be classified as a member of any of the following three groups: $X_{illegal}$, the set of illegal nodes, which corresponds to the cases when the cat and the mouse are both in the same room; X_m, the set of legal and marked nodes, which corresponds to the case when the cat is in Room 2 and the mouse in Room 4 (state 24 in G); and $X_{transient}$, the legal but unmarked nodes, which is made up of the rest of the nodes. (Observe that in this problem, the specification language K is completely characterized by a specification of the illegal states of G, the shuffle of G_c and G_m.)

Figures 2 and 3 show the first three variable lookahead windows expanded

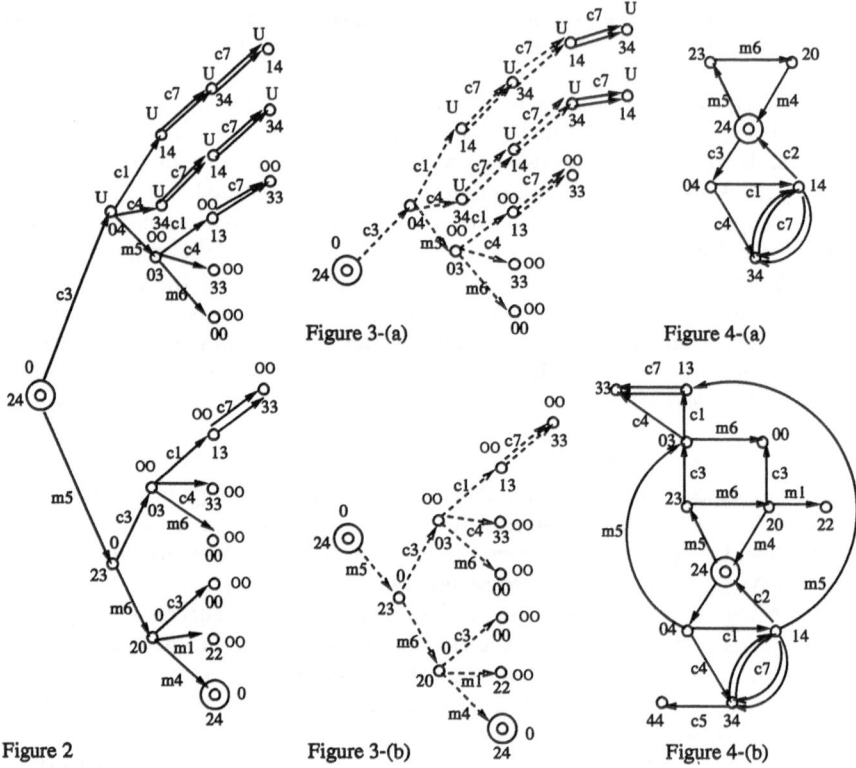

Figure 3-(a)

Figure 4-(a)

Figure 2

Figure 3-(b)

Figure 4-(b)

by the VLP algorithm with the undecided attitude, V_{vu}^N, when the window size $N = 4$. We use double lines to denote uncontrollable events, and dashed lines to denote computational results from previous steps. At the initial state 24, as indicated in figure 1, in order to derive the control action, only 22 nodes are expanded out of the 78 nodes in the complete 4-level tree at state 24. Between successive steps, only two additional nodes are expanded at state 04, as indicated in figure 3-(a), while no node is added at state 23, as indicated in figure 3-(b). Observe that for the sake of convenience, the nodes in figures 2 and 3 are labeled by their corresponding states of G; however, as was mentioned before, this is not part of the knowledge of the supervisor. We refer the reader to the discussion in the next section.

Table 1 lists the control actions derived by V_{vu}^N with different window sizes N, at all possible states (in G) that can occur in this LLP control scheme. These control actions are compared with the *valid* control actions, i.e., the ones that would be obtained by the standard off-line solution. For example, the control action at state 24 with window size $N = 4$ is equal to $\{c_3?, m_5\}$. The question mark ? denotes that the preceding event is an uncertain event, namely this event takes the system into a state with cost-to-go U. In other words, we cannot

	24	04	34	14	23	20	03	13
valid γ	c_3, m_5	c_1, c_4	c_7	c_2, c_7	m_6	m_4		
γ_{vu}^1	$c_3?, m_5?$	$c_1?, c_4?, m_5?$	$c_7?$	$c_2?, c_7?, m_5?$	$c_3?, m_6?$	$m_4?$	$c_1?$	$\emptyset$
γ_{vu}^2	$c_3?, m_5?$	$c_1?, c_4?, m_5?$	$c_7?$	$c_2, c_7?$	$c_3?, m_6?$	m_4	$\emptyset$	
γ_{vu}^3	$c_3?, m_5?$	$c_1?, c_4?$	$c_7?$	$c_2, c_7?$	m_6	m_4		
$\gamma_{vu}^N, N \geq 4$	$c_3?, m_5$	$c_1?, c_4?$	$c_7?$	$c_2, c_7?$	m_6	m_4		

Table 1: Control actions generated by V_{vu}^N.

decide unambiguously whether or not this event should be included in the final control action because of the uncertainty of the system behavior beyond the N-step boundary. An attitude has to be adopted to resolve this uncertainty, The $\emptyset$ at states 03 and 13 denotes that a *Run Time Error* occurs (cf. [2]); while a Run Time Error does not necessary imply a violation of the legality constraint, it at least, as in this exercise, means that an inner blocking happens. figure 4-(b) shows all the states that can be expanded by V_{vu}^1, the worst case of V_{vu}^N. This can be compared with figure 4-(a) which shows the recognizer of the optimal off-line solution in [5], the language $\overline{K^\uparrow}$.

As mentioned before, the control actions $\gamma_{l,cons}^N(s)$ ($\gamma_{l,optm}^N(s)$, respectively) derived by the conservative (optimistic, respectively) attitude can be obtained from that of $\gamma_{vu}^N(s)$ derived by the undecided attitude. In a nutshell, $\gamma_{l,cons}^N(s)$ ($\gamma_{l,optm}^N(s)$, respectively) just ignores (includes, respectively) the uncertain events present in $\gamma_{vu}^N(s)$. The closed loop behaviors that can be derived from table 1 when either the conservative or the optimistic attitude is adopted are as follows.

1. Conservative: When N is too small, the solution is not interesting:

$$L(G, \gamma_{l,cons}^N) = \epsilon, \ \forall N, 1 \leq N \leq 3$$

   where ϵ denotes the empty string. On the other hand, no matter how big N is, the closed loop behavior will never attain the supremal controllable sublanguage:

$$L(G, \gamma_{l,cons}^N) = (m_5 m_6 m_4)^* \subset \overline{K^\uparrow}, \ \forall N \geq 4.$$

2. Optimistic: When N is less than or equal to two, the system is likely to get into a blocking situation at either state 03 or state 13; however, the closed loop behavior will be the same as the optimal off-line solution if the window size is larger than or equal to three:

$$L(G, \gamma_{l,optm}^N) = \overline{K^\uparrow}, \ \forall N \geq 3.$$

4. DISCUSSION

Uncontrollable loop: Depending on the nature of the language generated by the system behavior, the VLP algorithm with no bound, V_v, can be computationally optimal provided the forward search procedure will always terminate without reaching the boundary. In this exercise, however, the required sufficient condition stated in Theorem 2 is not satisfied, and thus V_v need not terminate; this is due to the uncontrollable loop c_7c_7 connecting states 14 and 34. In addition, this uncontrollable loop also accounts for the problem that no matter how big the window size N is, the system behavior under the conservative attitude $L(G, \gamma_{l,cons}^N)$ can never attain $\overline{K^\uparrow}$.

The capability of identifying duplicate states: In the most general formulation of the LLP control scheme, the supervisor is not assumed to have the capability of identifying duplicate states (and thus loops) during the forward search procedure, as we do not want to impose any limitation on the way the system is modeled. (By duplicate states, we mean two nodes of the tree that correspond to the same system state.) However, in this exercise it is in fact feasible to assume that the supervisor does have such a capability, since a complete model of G is available. The implication is twofold. First, the problem caused by the uncontrollable loop can be eliminated if the VLP algorithm V_v is enhanced to make appropriate use of the additional state information (we shall discuss this issue in more detail in future work); with such information, algorithm V_v will always terminate (for finite-state systems). Second, due to the final value property of V_v, the decided cost-to-go, either 0 or ∞, at a given state should be the same whenever the state is visited during the forward search process, either within a step (e.g., in two different branches of the tree), or from step to step. Hence, if the supervisor can store the costs-to-go of the states the system has traversed, the on-line control action calculation will become necessary only when a new state is visited.

REFERENCES

[1] S. L. Chung. *Control of Discrete Event Systems Using Limited Lookahead Policies.* PhD thesis, Department of Electrical Engineering and Computer Science, University of Michigan, Ann Arbor, MI, July 1992.

[2] S. L. Chung, S. Lafortune, and F. Lin. Limited lookahead policies in supervisory control of discrete event systems. Technical Report CGR-70, College of Engineering Control Group Reports, University of Michigan, July 1991. To appear in *IEEE Trans. Automatic Control,* December 1992.

[3] S. L. Chung, S. Lafortune, and F. Lin. Recursive computation of limited lookahead supervisory controls for discrete event systems. Technical Report CGR-92-2, College of Engineering Control Group Reports, University of Michigan, March 1992.

[4] S. L. Chung, S. Lafortune, and F. Lin. Supervisory control using variable lookahead policies. Technical Report CGR-92-9, College of Engineering Control Group Reports, University of Michigan, July 1992.

[5] P. J. Ramadge and W. M. Wonham. Supervisory control of a class of discrete event systems. *SIAM J. Control and Optimization,* 25(1):206–230, January 1987.

[6] P. J. Ramadge and W. M. Wonham. The control of discrete event systems. *Proc. IEEE,* 77(1):81–98, January 1989.

SELECTED BIBLIOGRAPHY

SELECTED BIBLIOGRAPHY ON
DISCRETE EVENT SYSTEM CONTROL

The references listed below contain journal papers and books closely related to the modelling and control of discrete event systems. This bibliography, however, does not objectively represent the different research directions of the field. For instance, it does not contain references to stochastic systems, stochastic optimisation, operation research, and simulation. Only some references to perturbation analysis, scheduling, and min-max algebra are included. References to Petri nets are included only if they pertain to the latter topics. Nevertheless, the research area represented by the given references is the well-defined subarea of DES control at the logical level.

An extensive Petri nets bibliography collected by H. Plünnecke and W. Reisig has been published in *Advances in Petri Nets,* Lecture Notes in Computer Science No. 524, Springer Verlag, Berlin, 1991, pages 317–572. A "selected and annotated bibliography on perturbation analysis" by Y.C. Ho has appeared in *Discrete Event Systems: Models and Application,* Lecture Notes in Control and Information Sciences No. 103, Springer Verlag, Berlin, 1987, pages 217–224. Moreover, many interesting papers on DES control can be found in the proceedings of the IEEE Conference on Decision and Control, the American Control Conference, the IFAC World Congress, and the European Control Conference.

The items are alphabetically ordered according to the authors. A current version of a DES control BibTEX database including also conference papers, technical reports and other publications can be obtained via anonymous ftp from the server ifa.ethz.ch or via e-mail from the editors.

The editors

Aggarwal, S., Alonso, R. and Courcoubetis, C. (1987a). *Distributed Reachibility Analysis for Protocol Verification Environments*, Vol. 103 of Varaiya and Kurzhanski (1987), pp. 40–56. (IIASA Conference, Sopron, Hungary).

Aggarwal, S. and Gopinath, B. (1988). Special issue on tools for computer communication systems, *IEEE Trans. Software Eng.*, **14**(3): 277–404.

Aggarwal, S., Barbará, D. and Meth, K. (1987b). SPANNER: A tool for the specification, analysis, and evaluation of protocols, *IEEE Trans. Software Eng.*, **13**(12): 1218–1237.

Aggarwal, S., Barbará, D. and Meth, K. (1988). A software environment for the specification and analysis of problems of coordination and concurrency, *IEEE Trans. Software Eng.*, **14**(3): 280–290.

Akella, R. and Kumar, P. (1986). Optimal control of production rate in a failure prone manufacturing system, *IEEE Trans. Autom. Control*, **31**(2): 116–126.

Alur, R. and Dill, D. (1991). *The Theory of Timed Automata*, number 600 in *Lecture Notes in Computer Science Series*, Springer-Verlag, pp. 45–73. Proceedings of the REX workshop "Real Time: Theory in Practice".

Alur, R. and Henzinger, T. A. (1991). *Logics and Models of Real Time: A Survey*, number 600 in *Lecture Notes in Computer Science Series*, Springer-Verlag, pp. 74–106. Proceedings of the REX workshop "Real Time: Theory in Practice".

Anderson, D. (1988). Automated protocol implementation with RTAG, *IEEE Trans. Software Eng.*, **14**(3): 291–300.

Arbib, M. (1966). Automata theory and control theory – a rapprochement, *Automatica*, **3**: 161–189.

Arbib, M. and Manes, E. (1974). Foundations of system theory: Decomposable systems, *Automatica*, **10**: 285–302.

Arbib, M. and Ziegler, H. (1969). On the relevance of abstract algebra to control theory, *Automatica*, **5**: 589–606.

Archetti, F. and Sciomachen, A. (1987). *Representation, Analysis and Simulation of Manufacturing Systems by Petri Net Based Models*, Vol. 103 of Varaiya and Kurzhanski (1987), pp. 162–178. (IIASA Conference, Sopron, Hungary).

Aveyard, R. (1974). A boolean model for a class of discrete event systems, *IEEE Trans. Systems Man and Cybernetics*, **4**(3): 249–258.

Baccelli, F. and Makowski, A. (1989). Queueing models for systems with synchronization constraints, *Proc. of the IEEE*, **77**(1): 138–161.

Baccelli, F., Cohen, G. and Gaujal, B. (1991). Recursive equations and basic properties of timed petri nets, *Discrete Event Dynamic Systems: Theory and Applications*, **1**(4): 415–439.

Baccelli, F., Cohen, G., Olsder, G. and Quadrat, J. (1992). *Synchronization and Linearity*, Wiley.

Badrinath, B. and Ramamritham, K. (1988). Synchronizing transactions on objects, *IEEE Trans. Computers*, **37**(5): 541–547.

Balemi, S., Hoffmann, G., Gyugyi, P., Wong-Toi, H. and Franklin, G. (1993). Supervisory control of a rapid thermal multiprocessor, *Joint issue Automatica and IEEE Trans. Autom. Control on* Meeting the Challenge of Computer Science in the Industrial Applications of Control. to appear.

Banaszak, Z. and Krogh, B. (1990). Deadlock avoidance in flexible manufacturing systems with concurrently competing process flows, *IEEE Trans. Robotics and Automation*.

Benveniste, A. and Berry, G. (1991a). Special section: Another look at real-time programming, *Proc. of the IEEE*.

Benveniste, A. and Berry, G. (1991b). The synchronous approach to reactive and real-time systems, *IEEE Trans. Autom. Control*, **9**(79): 1270–1282.

Benveniste, A. and Guernic, P. L. (1990). Hybrid dynamical systems theory and the SIGNAL language, *IEEE Trans. Autom. Control*, **35**(5): 535–546.

Benveniste, A., Le Guernic, P. and Jacquemot, C. (1991). Synchronous programming with events and relations: the SIGNAL language and its semantics, *Science of Computer and Programming*, **16**: 103–149.

Berthelot, G. and Terrat, R. (1982). Petri nets theory for the correctness of protocols, *IEEE Trans. Communications*, **30**(12): 2497–2505.

Bhattacharya, P. and Ephremides, A. (1989). Optimal scheduling with strict deadlines, *IEEE Trans. Autom. Control*, **34**(7): 721–728.

Bochmann, D. and Posthoff, C. (1981). *Binäre Dynamische Systeme*, R. Oldenbourg Verlag München Wien.

Bochmann, G. (1985). Finite state description of communication protocols, *Computer Networks and ISDN Systems*, **2**: 361–372.

Bochmann, G., Cerny, E., Gagné, M., Jarda, C., Léveillé, A., Lacaille, C., Maksud, M., Raghunathan, K. and Sarikaya, B. (1982). Experience with formal specifications using and extended state transition model, *IEEE Trans. Communications*, **30**(12): 2506–2511.

Boel, R. and Schuppen, J. v. (1989). Distributed routing for load balancing, *Proc. of the IEEE*, **77**(1): 210–221.

Boukas, E. and Haurie, A. (1990). Manufacturing flow control and preventive maintenance: A stochastic control approach, *IEEE Trans. Autom. Control*, **35**(9): 1024–1031.

Brand, D. and Zafiropulo, P. (1983). On communicating finite-state machines, *J. ACM*, **30**(2): 323–342.

Brand, K. and Kopainsky, J. (1988). Principles and engineering of process control with Petri nets, *IEEE Trans. Autom. Control*, **33**(2): 138–149.

Brandt, R., Garg, V., Kumar, R., Lin, F., Marcus, S. and Wonham, W. (1990). Formulas for calculating supremal controllable and normal sublanguages, *System and Control Letters*, **15**(2): 111–117.

Brave, Y. and Heymann, M. (1990). Stabilization of discrete-event processes, *Int. J. Control*, **51**(5): 1101–1117.

Brave, Y. and Heymann, M. (1992). Optimal attraction in discrete event processes, *Information Sciences*.

Brookes, S., Hoare, C. and Roscoe, A. (1984). A theory of communicating sequential processes, *J. ACM*, **31**(3): 560–599.

Brooks, C., Cieslak, R. and Varaiya, P. (1990). A method for specifying, implementing and verifying media access control protocols, *IEEE Control System Magazine*, **10**(4): 87–94.

Browne, M., Clarke, E. and Grumberg, O. (1989). Reasoning about networks with many identical finite state processes, *Information and Computation*, **81**: 13–31.

Burns, A. and Wellings, A. (1990). *Real-Time Systems and their Programming Languages*, Addison-Wesley.

Buzacott, J. (1982). "Optimal" operating rules for automated manufacturing systems, *IEEE Trans. Autom. Control*, **27**(1): 80–86.

Cao, X. (1985). Convergence of parameter sensitivity estimates in a stochastic experiment, *IEEE Trans. Autom. Control*, **30**(9): 845–853.

Cao, X. (1989a). The predictability of discrete event systems, *IEEE Trans. Autom. Control*, **34**(11): 1168–1171.

Cao, X. (1989b). The static property of perturbed multiclass closed queuing network and decomposition, *IEEE Trans. Autom. Control*, **34**(2): 246–249.

Cao, X. and Ho, Y. (1987). Sensitivity analysis and optimization of throughput in a production line with blocking, *IEEE Trans. Autom. Control*, **32**(11): 959–967.

Cao, X.-R. and Ho, Y.-C. (1990). Models of discrete event systems, *IEEE Control System Magazine*, **10**(4): 69–76.

Cao, X.-R. and Ma, D.-J. (1991). Event rates and aggregation in hierarchical systems, *Discrete Event Dynamic Systems: Theory and Applications*, **1**(3): 271–288.

Cassandras, C. (1990). Toward a control theory for discrete event systems, *IEEE Control System Magazine*, **10**(4): 66–68.

Cassandras, C. and Strickland, S. (1989a). Observable augmented systems for sensitivity analysis of Markov and semi-Markov processes, *IEEE Trans. Autom. Control*, **34**(10): 1026–1037.

Cassandras, C. and Strickland, S. (1989b). On line sensitivity analysis of Markov chains, *IEEE Trans. Autom. Control*, **34**(1): 76–86.

Cassandras, C. and Strickland, S. (1989c). Sample path properties of timed discrete event systems, *Proc. of the IEEE*, **77**(1): 59–71.

Chase, C. and Ramadge, P. (1992). On real-time scheduling policies for flexible manufacturing systems, *IEEE Trans. Autom. Control*, **37**(4): 491–496.

Chen, E. and Lafortune, S. (1991a). Dealing with blocking in supervisory control of discrete event systems, *IEEE Trans. Autom. Control*, **36**(6): 724–735.

Chen, E. and Lafortune, S. (1991b). On nonconflicting languages that arise in supervisory control of discrete event systems, *System and Control Letters*, **17**: 105–113.

Cho, H. and Marcus, S. (1989a). On supremal languages of classes of sublanguages that arise in supervisor synthesis problems with partial observation, *Mathematics of Control, Signals and Systems*, **2**(2): 47–69.

Cho, H. and Marcus, S. (1989b). Supremal and maximal sublanguages arising in supervisor synthesis problems with partial observations, *Math. Systems Theory*, **22**: 177–211.

Chong, E. and Ramadge, P. (1991). Convergence of infinitesimal optimizationn algorithms using infinitesimal perturbation analysis estimates, *Discrete Event Dynamic Systems: Theory and Applications*, **1**(4): 339–372.

Choueka, Y. (1974). Theories of automata on ω-tapes: A simplified approach, *Journal of Computer and System Sciences*, **8**: 117–141.

Cieslak, R. and Varaiya, P. (1990). Undecidability results for deterministic communicating sequential processes, *IEEE Trans. Autom. Control*, **35**(9): 1032–1039.

Cieslak, R. and Varaiya, P. (1991). Simulating finitely recursive processes, *Discrete Event Dynamic Systems: Theory and Applications*, **1**(4): 373–392.

Cieslak, R., Desclaux, C., Fawaz, A. and Varaiya, P. (1988). Supervisory control of discrete-event processes with partial observations, *IEEE Trans. Autom. Control*, **33**(3): 249–260.

Cohen, G., Dubois, D., Quadrat, J. and Viot, M. (1985). A linear-system-theoretic view of discrete-event processes and its use for performance evaluation in manufacturing, *IEEE Trans. Autom. Control*, **30**(3): 210–220.

Cohen, G., Moller, P., Quadrat, J. and Viot, M. (1989). Algebraic tools for the performance evaluation of discrete event systems, *Proc. of the IEEE*, **77**(1): 39–58.

Collective (1987). Challenges to control: A collective view, *IEEE Trans. Autom. Control*, **32**(4): 275–285.

Concepcion, A. and Zeigler, B. (1988). DEVS formalism: A formalism for hierarchical model development, *IEEE Trans. Software Eng.*, **14**(2): 228–241.

Cuninghame-Green, R. (1989). Discrete events and max-algebra, *CWI Quarterly*, **2**(3): 257–258.

David, R. and Alla, H. (1992). *Petri Nets & Grafcet: Tools for modelling discrete event systems*, Prentice Hall.

Dayhoff, J. and Atherton, R. (1987). A model for wafer fabrication dynamics in integrated circuit manufacturing, *IEEE Trans. Systems Man and Cybernetics*, **17**(1): 91–100.

Delchamps, D. (1990). Stabilizing a linear system with quantized state feedback, *IEEE Trans. Autom. Control*, **35**(8): 916–924.

Denardo, E. (1982). *Dynamic Programming: Models and Applications*, Prentice-Hall Inc., Englewood Cliffs.

Denham, M. (1988). *A Petri Net Approach to the Control of Discrete-Event Systems*, Vol. 47 of Denham and Laub (1988), pp. 191–214.

Denham, M. and Laub, A. (eds) (1988). *Advanced Computing Concepts and Techniques in Control Engineering*, Vol. 47 of *Computer and Systems Sciences*, Springer Verlag.

DeNicola, R. (1987). Extensional equivalences for transition systems, *Acta Informatica*, **24**: 211–237.

Desrochers, A. (1990). *Modeling and Control of Automated Manufacturing Systems*, IEEE Computer Society Press, Washington, DC.

Diaz, M. (1982). Modeling and analysis of communication protocols using Petri net based models, *Computer Networks and ISDN Systems*, **6**: 439–441.

Dill, D. (1989). *Trace Theory for Automatic Hierarchical Verification of Speed-Independent Circuits*, ACM Distinguished Dissertations, The MIT Press, Cambridge, Massachussetts.

Du, Y. and Wang, S. (1988). Control of discrete-event systems with minimal switchings, *Int. J. Control*, **48**(3): 981–991.

Du, Y. and Wang, S. (1989). Translation of output constraint into event constraint in the control of discrete event systems, *Int. J. Control*, **50**(6): 2635–2644.

Ephremides, A. and Verdú, S. (1989). Control and optimization methods in communication network problems, *IEEE Trans. Autom. Control*, **34**(9): 930–942.

Ezzine, J. and Haddad, A. (1989). Error bounds in the averaging of hybrid systems, *IEEE Trans. Autom. Control*, **34**(11): 1188–1191.

Fendick, K. and Whitt, W. (1989). Measurements and approximations to describe the offered traffic and predict the average workload in a single-server queue, *Proc. of the IEEE*, **77**(1): 171–194.

Gershwin, S. (1987). *A Hierarchical Framework for Discrete Event Scheduling in Manufacturing Systems*, Vol. 103 of Varaiya and Kurzhanski (1987), pp. 197–216. (IIASA Conference, Sopron, Hungary).

Gershwin, S. (1989). Hierarchical flow control: A framework for scheduling and planning discrete events in manufacturing systems, *Proc. of the IEEE*, **77**(1): 195–209.

Glasserman, P. and Gong, W. (1990). Smoothed perturbation analysis for a class of discrete-event systems, *IEEE Trans. Autom. Control*, **35**(11): 1218–1230.

Glasserman, P. and Yao, D. (1991). Algebraic structure of some stochastic discrete event systems, with applications, *Discrete Event Dynamic Systems: Theory and Applications*, **1**(1): 7–36.

Glynn, P. (1989). A GSMP formalism for discrete event systems, *Proc. of the IEEE*, **77**(1): 14–23.

Golaszewski, C. and Ramadge, P. (1988). *Supervisory Control of Discrete Event Processes with Arbitrary Controls*, Vol. 47 of Denham and Laub (1988), pp. 459–469.

Golaszewski, C. and Ramadge, P. (1989). The complexity of some reachability problems for a system on a finite group, *System and Control Letters*, **12**: 431–435.

Gong, W. and Cassandras, C. (1991). Perturbation analysis of a multiclass queueing system with admission control, *IEEE Trans. Autom. Control*, **36**(6): 707–723.

Gong, W. and Ho, Y. (1987). Smoothed perturbation analysis of discrete event dynamic systems, *IEEE Trans. Autom. Control*, **32**(11): 858–866.

Habermann, A. (1969). Prevention of systems deadlocks, *Communications of the ACM*, **12**(7): 373–385.

Harel, D. (1987). Statecharts: A visual formalism for complex system, *Science of Computer Programming*, **8**(3): 231–274.

Harel, D. (1992). Biting the silver bullet: Toward a brighter future for software development, *IEEECM*, **25**(1): 8–20.

Hatono, I., Yamagata, K. and Tamura, H. (1991). Modeling and on-line scheduling of flexible manufacturing systems using stochastic Petri nets, *IEEE Trans. Software Eng.*, **17**(2): 126–132.

Hennessy, M. and Milner, R. (1985). Algebraic laws for nondeterminism and concurrency, *J. ACM*, **32**(1): 137–161.

Heymann, M. (1990). Concurrency and discrete event control, *IEEE Control System Magazine*, **10**(4): 103–112.

Hillion, H. and Proth, J. (1989). Performance evaluation of job-shop systems using timed event-graphs, *IEEE Trans. Autom. Control*, **34**(1): 3–9.

Ho, Y. (1985). On the perturbation analysis of discrete-event dynamic systems, *Journal of Optimization Theory and Applications*, **46**(4): 535–545.

Ho, Y. (1987a). Performance evaluation and perturbation analysis of discrete event dynamic systems, *IEEE Trans. Autom. Control*, **32**(7): 563–572.

Ho, Y. (1987b). *A Selected and Annotated Bibliography on Perturbation Analysis*, Vol. 103 of Varaiya and Kurzhanski (1987), pp. 217–224. (IIASA Conference, Sopron, Hungary).

Ho, Y. (1989a). Dynamics of discrete event systems, *Proc. of the IEEE*, **77**(1): 3–6. Introduction.

Ho, Y. and Cao, X. (1985). Perturbation analysis and optimization of queueing networks, *Journal of Optimization Theory and Applications*, **40**: 559–582.

Ho, Y. and Cao, X. (1991). *Perturbation Analysis of Discrete Event Dynamic Systems*, Kluwer Academic Publishers.

Ho, Y. and Cassandras, C. (1983). A new approach to the analysis of discrete event dynamic systems, *Automatica*, **19**: 149–167.

Ho, Y. and Li, S. (1988). Extensions of infinitesimal perturbation analysis, *IEEE Trans. Autom. Control*, **33**(5): 427–438.

Ho, Y. C. (1989b). Special issue on the dynamics of discrete event systems, *Proc. of the IEEE*.

Hoare, C. (1978). Communicating sequential processes, *Communications of the ACM*, **21**(8): 666–677.

Hoare, C. (1985). *Communicating Sequential Processes*, International Series in Computer Science, Prentice-Hall International, Englewood Cliffs, NJ.

Holloway, L. and Krogh, B. (1990). Synthesis of feedback control logic for a class of controlled Petri nets, *IEEE Trans. Autom. Control*, **35**(5): 514–523.

Hopcroft, J. E. and Ullman, J. D. (1969). *Formal Languages and their Relation to Automata*, Addison-Wesley Series in Computer Science, Addison-Wesley.

Hopcroft, J. E. and Ullman, J. D. (1979). *Introduction to Automata Theory, Languages and Computation*, Addison-Wesley Series in Computer Science, Addison-Wesley.

Houben, G., Dietz, J. and van Hee, K. (1987). *The SMARTIE Framework for Modelling Discrete Dynamic Systems*, Vol. 103 of Varaiya and Kurzhanski (1987), pp. 179–196. (IIASA Conference, Sopron, Hungary).

Hyung, L. and Lee, K. (1991). Finding the shortest path in railroad systems using the super-network, *Int. J. Systems Science*, **22**(7): 1307–1314.

Ichikawa, A. and Hiraishi, K. (1987). *Analysis and Control of Discrete Event Systems Represented by Petri Nets*, Vol. 103 of Varaiya and Kurzhanski (1987), pp. 115–134. (IIASA Conference, Sopron, Hungary).

İnan, K. (1992). An algebraic approach to supervisory control, *Mathematics of Control, Signals and Systems*, **2**(5): 151–164.

Inan, K. and Varaiya, P. (1987). *Finitely Recursive Processes*, Vol. 103 of Varaiya and Kurzhanski (1987), pp. 1–18. (IIASA Conference, Sopron, Hungary).

Inan, K. and Varaiya, P. (1988). Finitely recursive process models for discrete event systems, *IEEE Trans. Autom. Control*, **33**(7): 626–639.

Inan, K. and Varaiya, P. (1989). Algebras of discrete event models, *Proc. of the IEEE*, **77**(1): 24–38.

Javor, A. (1975). An approach to the modelling of uncertainties in the simulation of quasideterministic discrete event systems., *Problems of Control and Information Theory*, **4**(3): 219–229.

Kahn, G. (ed.) (1987). *Functional Programming Languages and Computer Architecture*, Vol. 274 of *Lecture Notes in Computer Science*, Springer Verlag, Berlin.

Kamath, M. and Sanders, J. (1991). Modeling operator/workstation interference in asynchronous automatic assembly systems, *Discrete Event Dynamic Systems: Theory and Applications*, **1**(1): 93–124.

Kavi, K., Buckles, B. and Bhat, U. (1987). Isomorphisms between Petri nets and dataflow graphs, *IEEE Trans. Software Eng.*, **13**(10): 1127–1134.

Kim, K.-H. and Hwang, C.-S. (1990). Research trends and the theory of distributed simulation, *Korea Information Science Society Review*, **8**(1): 56–63.

Kodama, S. and Kumagai, S. (1985). Discrete event system-net model approach, *Journal of the Society of Instrument and Control Engineers*, **24**(7): 623–632.

Kouikoglou, V. and Phillis, Y. (1991). An exact discrete-event model and control policies for production linies with buffers, *IEEE Trans. Autom. Control*, **36**(5): 515–527.

Kozák, P. (1992a). Discrete events and general systems theory, *Int. J. Systems Science*, **23**(9): 1403–1422.

Kozák, P. (1992b). On feedback controllers, *Int. J. Systems Science*, **23**(9): 1423–1431.

Kramer, J., Magee, J. and Sloman, M. (1984). A software architecture for distributed computer control systems, *Automatica*, **20**(1): 93–102.

Kramer, J., Magee, J. and Sloman, M. (1988). *An Overview of Distributed System Construction Using CONIC*, Vol. 47 of Denham and Laub (1988), pp. 237–255.

Krogh, B. and Feng, D. (1989). Dynamic generation of subgoals for autonomous mobile robots using local feedback information, *IEEE Trans. Autom. Control*, **34**(5): 483–493.

Kuipers, B. (1989). Qualitative reasoning: Modelling and simulation with incomplete knowledge, *Automatica*, **25**(4): 571–585.

Kumar, R., Garg, V. and Marcus, S. (1991). On controllability and normality of discrete event dynamical systems, *System and Control Letters*, **17**: 157–168.

Kurshan, R. (1987). *Reducibility in Analysis of Coordination*, Vol. 103 of Varaiya and Kurzhanski (1987), pp. 19–39. (IIASA Conference, Sopron, Hungary).

Lafortune, S. (1988). Modeling and analysis of transaction execution in database systems, *IEEE Trans. Autom. Control*, **33**(5): 439–447.

Lafortune, S. and Chen, E. (1990). The infimal closed controllable superlanguage and its application in supervisory control, *IEEE Trans. Autom. Control*, **35**(4): 398–405.

Lafortune, S. and Lin, F. (1991). On tolerable and desirable behavior in supervisory control of discrete event systems, *Discrete Event Dynamic Systems: Theory and Applications*, **1**(1): 61–92.

Lafortune, S. and Wong, E. (1986). A state transition model for distributed query processing, *ACM Trans. Database Systems*, **11**(3): 294–322.

Lafortune, S. and Yoo, H. (1990). Some results on Petri net languages, *IEEE Trans. Autom. Control*, **35**(4): 482–485.

Larsen, F. and Evans, F. (1988). *Structural Design of Decentralized Control Systems*, Vol. 47 of Denham and Laub (1988), pp. 257–288.

L'Ecuyer, P. and Haurie, A. (1988). Discrete event dynamic programming with simultaneous events, *Mathematics of Operations Research*, **13**(1): 152–163.

Lee, K.-H. and Favrel, J. (1985). Hierarchical reduction method for analysis and decomposition of Petri nets, *IEEE Trans. Systems Man and Cybernetics*, **15**(2): 272–280.

Lee-Kwang, H., Favrel, J. and Oh, G.-R. (1987). Hierarchical decomposition of Petri net languages, *IEEE Trans. Systems Man and Cybernetics*, **17**(5): 877–878.

Lee, T. and Lai, M.-Y. (1988). A relational algebraic approach to protocol verification, *IEEE Trans. Software Eng.*, **14**(2): 184–193.

Leeming, A. (1981). A comparison of some discrete event simulation languages, *Simuletter*, **12**(1–4): 9–16.

Leung, Y.-T. and Suri, R. (1990). Performance evaluation of discrete manufacturing systems, *IEEE Control System Magazine*, **10**(4): 77–86.

Levi, S.-T. and Agrawala, A. (1990). *Real-Time System Design*, McGraw-Hill, New York.

Levis, A. (1984). Information processing and decision-making organizations: A mathematical description, *Large Scale Systems*, **7**: 151–163.

Li, Y. and Wonham, W. (1988). On supervisory control of real-time discrete event systems, *Information Sciences*, **46**(3): 159–183.

Lin, F. (1991). Control of large scale discrete event systems: Task allocation and coordination, *System and Control Letters*, **17**: 169–175.

Lin, F. and Wonham, W. (1988a). Decentralized supervisory control of discrete-event systems, *Information Sciences*, **44**(3): 199–224.

Lin, F. and Wonham, W. (1988b). On observability of discrete-event systems, *Information Sciences*, **44**(3): 173–198.

Lin, F. and Wonham, W. (1990). Decentralized control and coordination of discrete-event systems with partial observation, *IEEE Trans. Autom. Control*, **35**(12): 1330–1337.

Lin, F., Vaz, A. and Wonham, W. (1988). Supervisor specification and synthesis for discrete event systems, *Int. J. Control*, **48**(1): 321–332.

Lingarkar, R., Liu, L., Elbestavwi, M. and Sinha, N. (1990). Knowledge-based adaptive computer control in manufacturing systems: A case study, *IEEE Trans. Systems Man and Cybernetics*, **20**(3): 606–618.

Looney, C. and Alfize, A. (1987). Logical controls via boolean rule matrix transformations, *IEEE Trans. Systems Man and Cybernetics*, **17**(6): 1077–1082.

Lynch, N. and Tuttle, M. (1989). An introduction to input/output automata, *CWI Quarterly*, **2**(3): 219–246.

Magott, J. (1984). Performance evaluation of concurrent systems using Petri nets, *Information Processing Letters*, **18**(1): 7–13.

Maimon, O. and Tadmor, G. (1988). Model-mased low-level control in flexible manufacturing systems, *Robotics & Computer-Integrated Manufacturing*, **4**(3/4): 423–428.

Maler, O., Manna, Z. and Pnueli, A. (1991). *From Timed to Hybrid Systems*, number 600 in *Lecture Notes in Computer Science Series*, Springer-Verlag, pp. 447–484. Proceedings of the REX workshop "Real Time: Theory in Practice".

Manna, Z. and Pnueli, A. (1992). *The temporal logic of reactive and concurrent systems*, Springer-Verlag.

McIntyre, M. (1987). Some direct interconnections between formal language theory and discrete-time linear system theory, *IEEE Trans. Circuits and Systems*, **34**(9): 1102–1107.

Merlin, P. and Bochmann, G. (1983). On the construction of submodule specifications and communication protocols, *ACM Trans. Programming Languages and Systems*, **5**(1): 1–25.

Milner, R. (1980). *A Calculus of Communicating Systems*, Vol. 92 of *LNCS*, Springer-Verlag, New York.

Milner, R. (1989). *Communication and Concurrency*, Prentice Hall, New York.

Mowshowitz, A. (1968). Entropy and the complexity of graphs, *Bulletin Math. Biophys.*, **30**: 175–204.

Muhanna, W. (1991). Composite programs: Hierarchical construction, circularity, and deadlocks, *IEEE Trans. Software Eng.*, **17**(4): 320–333.

Murata, T. (1977a). Petri nets, marked graphs, and circuit-system theory, *IEEE Trans. Circuits and Systems*.

Murata, T. (1977b). State equations, controllability, and maximal matchings of Petri nets, *IEEE Trans. Autom. Control*, **22**(3): 412–416.

Murata, T. (1980). Synthesis of decision-free concurrent systems for prescribed resources and performance, *IEEE Trans. Software Eng.*, **6**(6): 525–530.

Murata, T. (1989). Petri nets: Properties, analysis, and applications, *Proc. of the IEEE*, **77**(4): 541–580.

Naylor, A. and Volz, R. (1987). Design of integrated manufacturing system control software, *IEEE Trans. Systems Man and Cybernetics*, **17**(6): 881–897.

Olderog, E.-R. (1991). *Nets, Terms and Formulas*, Vol. 23 of *Cambridge Tracts in Theoretical Computer Science*, Cambridge University Press.

Olsder, G. (1991). Eigenvalues of dynamic max-min systems, *Discrete Event Dynamic Systems: Theory and Applications*, **1**(2): 177–207.

Olsder, G. and de Vries, R. (1987). *On an Analogy of Minimal Realizations in Conventional and Discrete-Event Dynamic Systems*, Vol. 103 of Varaiya and Kurzhanski (1987), pp. 149–161. (IIASA Conference, Sopron, Hungary).

Olsder, G., Resing, J., de Vries, R., Keane, M. and Hooghiemstra, G. (1990). Discrete event systems with stochastic processing times, *IEEE Trans. Autom. Control*, **35**(3): 299–302.

Oommen, B. and Hansen, E. (1984). The asymptotic optimality of discretized linear reward-inaction learning automata, *IEEE Trans. Systems Man and Cybernetics*, **14**(3): 542–545.

Ostroff, J. (1988). *Temporal Logic and Extended State Machines in Discrete Event Systems*, Vol. 47 of Denham and Laub (1988), pp. 215–236.

Ostroff, J. (1989). *Temporal Logic for Real-Time Systems*, RSP, Research Studies Press / Wiley. Taunton, UK.

Ostroff, J. (1990). A logic for real-time discrete event processes, *IEEE Control System Magazine*, **10**(4): 95–102.

Ostroff, J. and Wonham, W. (1990). A framework for real-time discrete event control, *IEEE Trans. Autom. Control*, **35**(4): 386–397.

Özsu, M., Wong, K. and Koon, T. (1988). System modeling and analysis using Petri nets, *Syst. Anal. Model. Simul.*, **5**(1): 3–25.

Özveren, C. and Willsky, A. (1990). Observability of discrete event systems, *IEEE Trans. Autom. Control*, **35**(7): 797–806.

Özveren, C. and Willsky, A. (1991). Output stabilizability of discrete event dynamic systems, *IEEE Trans. Autom. Control*, **36**(8): 925–935.

Özveren, C. and Willsky, A. (1992a). Aggregation and multi-level control in discrete event systems, *Automatica*, **28**(3): 565–578.

Özveren, C. and Willsky, A. (1992b). Invertibility of discrete event dynamic systems, *Mathematics of Control, Signals and Systems*, **5**(4): 365–390.

Özveren, C., Willsky, A. and Antsaklis, P. (1991). Stability and stabilizality of discrete event dynamic systems, *J. ACM*, **38**(3): 730–752.

Passino, K. and Antsaklis, P. (1991). Event rates and aggregation in hierarchical systems, *Discrete Event Dynamic Systems: Theory and Applications*, **1**(3): 271–288.

Perdu, D. and Levis, A. (1991). A Petri net model for evaluation of expert systems in organizations, *Automatica*, **27**(2): 225–237.

Perkins, J. and Kumar, P. (1989). Stable, distributed, real-time scheduling of flexible manufacturing/assembly/disassembly systems, *IEEE Trans. Autom. Control*, **34**(2): 139–148.

Peterson, J. (1977). Petri nets, *ACM Computing Surveys*, **9**(3): 223–252.

Peterson, J. (1981). *Petri Net Theory and the Modelling of Systems*, Prentice-Hall, Englewood Cliffs.

Pnueli, A. (1986). *Applications of Temporal Logic to the Specification and Verification of Reactive Systems: A Survey and Current Trends*, number 224 in *Lecture Notes in Computer Science Series*, Springer-Verlag, pp. 510–584.

Ramadge, P. (1987). *Supervisory Control of Discrete Event Systems: A Survey and Some New Results*, Vol. 103 of Varaiya and Kurzhanski (1987), pp. 69–80. (IIASA Conference, Sopron, Hungary).

Ramadge, P. (1988). *The Complexity of Some Basic Control Problems for Discrete Event Systems*, Vol. 47 of Denham and Laub (1988), pp. 171–190.

Ramadge, P. (1989). Some tractable supervisory control problems for discrete-event systems modeled by Büchi automata, *IEEE Trans. Autom. Control*, **34**(1): 10–19.

Ramadge, P. (1990). On the periodicity of symbolic observations of piecewise smooth discrete-time systems, *IEEE Trans. Autom. Control*, **35**(7): 807–813.

Ramadge, P. and Wonham, W. (1987a). Modular feedback logic for discrete event systems, *SIAM J. Control Optim.*, **25**(5): 1202–1218.

Ramadge, P. and Wonham, W. (1987b). Supervisory control of a class of discrete event processes, *SIAM J. Control Optim.*, **25**(1): 206–230.

Ramadge, P. and Wonham, W. (1989). The control of discrete event systems, *Proc. of the IEEE*, **77**(1): 81–98.

Rea, K. and Johnston, R. (1987). Automated analysis of discrete communication behavior, *IEEE Trans. Software Eng.*, **13**(10): 1115–1126.

Reisig, W. (1985). *Petri Nets. An Introduction*, Vol. 4 of *Monographs on Theoretical Computer Science*, Springer-Verlag.

Righter, R. and Walrand, J. (1989). Distributed simulation of discrete event systems, *Proc. of the IEEE*, **77**(1): 99–113.

Riordon, J. (1969a). An adaptive automaton controller for discrete-time Markov processes, *Automatica*, **5**: 721–730.

Riordon, J. (1969b). Optimal feedback characteristics from stochastic automaton models, *IEEE Trans. Autom. Control*, **14**: 89–92.

Rosenberger, F., Molnar, C., Chaney, T. and Fang, T. (1988). II-modules: Internally clocked delay-insensitive modules, *IEEE Trans. Computers*, **37**(9): 1005–1018.

Roszkowska, E. and Banaszak, Z. (1991). *Problems of Deadlock Handling in Pipeline Processes*, Computer and Information Sciences VI, Elsevier Science Publisher.

Rudie, K. and Wonham, W. (1990). The infimal prefix-closed and observable superlanguage of a given language, *System and Control Letters*, **15**(5): 361–371.

Rudie, K. and Wonham, W. (1992). Think globally, act locally: Decentralized supervisory control, *IEEE Trans. Autom. Control*, **37**(11): 1692–1708.

Sabnani, K. (1988). An algorithmic technique for protocol verification, *IEEE Trans. Communications*, **36**(8): 924–931.

Sajkowski, M. (1987). *Protocol Verification Using Discrete-Event Models*, Vol. 103 of Varaiya and Kurzhanski (1987), pp. 100–114. (IIASA Conference, Sopron, Hungary).

Schumacher, J. (1989). Discrete events: Perspectives from system theory, *CWI Quarterly*, **2**(2): 131–146.

Schwartz, R. and Melliar-Smith, P. (1982). From state machines to temporal logic: Specification methods for protocol standards, *IEEE Trans. Communications*, **30**(12): 2486–2496.

Seiche, W. (1991). Control synthesis based on a graph-theoretical Petri net analysis, *Int. J. of Microelectronics and Reliability*, **31**(4): 563–575.

Sengupta, R. and Lafortune, S. (1991). A graph-theoretical optimal control problem for terminating discrete event processes, *Discrete Event Dynamic Systems: Theory and Applications*, **2**(1): 139–172.

Shanthikumar, J. and Yao, D. (1989). Second order stochastic properties of queueing systems, *Proc. of the IEEE*, **77**(1): 162–170.

Sharifnia, A., Caramanis, M. and Gershwin, S. (1991). Dynamic setup scheduling and flow control in manufacturing systems, *Discrete Event Dynamic Systems: Theory and Applications*, **1**(2): 149–175.

Shreve, S. and Bertsekas, D. (1979). Universally measurable policies in dynamic programming, *Mathematics of Operations Research*, **4**(1): 15–30.

Smedinga, R. (1987). *Using Trace Theory to Model Discrete Events*, Vol. 103 of Varaiya and Kurzhanski (1987), pp. 81–99. (IIASA Conference, Sopron, Hungary).

Smedinga, R. (1993). Locked discrete event systems: how to model and how to unlock, *Discrete Event Dynamic Systems: Theory and Applications*.

Sparaggis, P. and Cassandras, C. (1991). Monotocity properties of cost functions in queueing networks, *Discrete Event Dynamic Systems: Theory and Applications*, **1**(3): 315–327.

Sreenivas, R. and Krogh, B. (1991). On condition/event systems with discrete state realizations, *Discrete Event Dynamic Systems: Theory and Applications*, **1**(2): 209–235.

Sreenivas, R. and Krogh, B. (1992). On Petri nets models of infinite state supervisors, *IEEE Trans. Autom. Control*, **37**(2): 274–277.

Stankovic, J. A. and Ramamritham, K. (eds) (1988). *Tutorial: Hard Real-Time Systems*, IEEE Computer Society Press, Washington, D.C.

Subrahmanian, E. and Canon, R. (1981). A generator program for models of discrete-event systems, *Simulation*, **36**(3): 93–101.

Sunshine, C. (1982). Special issue on protocol specification, testing, and verification, *IEEE Trans. Communications*.

Suri, R. (1989). Perturbation analysis: The state of the art and research issues explained via the G/G/1 queue, *Proc. of the IEEE*, **77**(1): 114–137.

Suzuki, T., Shatz, S. and Murata, T. (1990). A protocol modeling and verification approach based on a specification language and Petri nets, *IEEE Trans. Software Eng.*, **16**(5): 523–536.

Tabak, D. and Levis, A. (1985). Petri net representation of decision models, *IEEE Trans. Systems Man and Cybernetics*, **15**(6): 812–818.

Tadmor, G. and Maimon, O. (1989). Control of large discrete event systems: Constructive algorithms, *IEEE Trans. Autom. Control*, **34**(11): 1164–1168.

Thistle, J. and Wonham, W. (1986). Control problems in a temporal logic framework, *Int. J. Control*, **44**(4): 943–976.

Tsitsiklis, J. (1989). On the control of discrete event dynamical systems, *Mathematics of Control, Signals and Systems*, **2**(2): 95–107.

Ushio, T. (1989a). Controllability and control-invariance in discrete-event systems, *Int. J. Control*, **50**(4): 1507–1515.

Ushio, T. (1989b). Controllable languages and predicates in discrete-event systems, *Transactions of the Institute of Electronics, Information and Communication Engineers A*, **J72A**(12): 2048–2051.

Ushio, T. (1989c). Modular feedbacks of discrete event systems with arbitrary control patterns, *Transactions of the Institute of Electronics, Information and Communication Engineers A*, **J72A**(6): 938–944.

Ushio, T. (1990a). Feedback logic for discrete event systems with arbitrary control patterns, *Int. J. Control*, **52**(1): 159–174.

Ushio, T. (1990b). Fixpoint characterization of controllability in discrete event systems with nondeterministic supervisors, *Transactions of the Institute of Electronics, Information and Communication Engineers A*, **J73A**(3): 642–644.

Ushio, T. (1990c). Maximally permissive feedback and modular control synthesis in Petri nets with external input places, *IEEE Trans. Autom. Control*, **35**(7): 844–848.

Ushio, T. (1990d). Modular feedbacks of discrete event systems with arbitrary control patterns, *Electronics and Communications in Japan, Part 3 (Fundamental Electronic Science)*, **73**(4): 99–106.

Van Breusegem, V., Campion, G. and Bastin, G. (1991). Traffic modelling and state feedback control for metro rail lines based on linear discrete-event state-space models, *IEEE Trans. Autom. Control*, **36**(7): 770–784.

Van de Snepscheut, J. (1985). *Trace Theory and VLSI Design*, Vol. 200 of *LNCS*, Springer-Verlag, New York.

Van der Rhee, F., Van Nauta Lemke, H. and Dukman, J. (1990). Knowledge based fuzzy control of systems, *IEEE Trans. Autom. Control*, **35**(2): 148–155.

Varaiya, P. (1987). *Preface*, Vol. 103 of Varaiya and Kurzhanski (1987), pp. iv–vii. (IIASA Conference, Sopron, Hungary).

Varaiya, P. and Kurzhanski, H. (eds) (1987). *Discrete Event Systems: Models and Application*, Vol. 103 of *Lecture Notes in Control and Information Sciences*, Springer Verlag, Berlin, Germany. (IIASA Conference, Sopron, Hungary).

Varaiya, P., Walrand, J. and Buyukkoc, C. (1985). Extensions of the multi-armed bandit problem: The discounted case, *IEEE Trans. Autom. Control*, **30**(5): 426–439.

Vaz, A. and Wonham, W. (1986). On supervisor reduction in discrete event systems, *Int. J. Control*, **44**(2): 475–491.

Wang, Z.-Q., Song, W.-Z. and Feng, C.-B. (1991). Full-state perturbation ananlysis of discrete event dynamic systems, *Discrete Event Dynamic Systems: Theory and Applications*, **1**(3): 249–270.

Wardi, Y. and Hu, J. (1991). Strong consistency of infinitesimal perturbation analysis for tandem queueing networks, *Discrete Event Dynamic Systems: Theory and Applications*, **1**(1): 37–60.

Wardi, Y., Gong, W.-B., Cassandras, C. and Kallmes, M. (1991). Smoothed perturbation analysis for a class of piecewise constant sample performance functions, *Discrete Event Dynamic Systems: Theory and Applications*, **1**(4): 393–414.

Whitney, D. (1969). State space models of remote manipulation tasks, *IEEE Trans. Autom. Control*, **14**(6): 617–623.

Willson, R. and Krogh, B. (1990). Petri net tools for the specification and analysis of discrete controllers, *IEEE Trans. Software Eng.*, **16**(1): 39–50.

Wirth, N. (1977). Toward a discipline of real-time programming, *Communications of the ACM*, **20**(8): 577–583.

Wolper, P. (1983). Temporal logic can be more expressive, *Information and Control*, **56**: 72–99.

Wonham, W. (1987). Some remarks on control and computer science, *IEEE Control System Magazine*, **7**(2): 9–10.

Wonham, W. (1988). *A Control Theory for Discrete Event Systems*, Vol. 47 of Denham and Laub (1988), pp. 129–169.

Wonham, W. and Ramadge, P. (1987). On the supremal controllable sublanguage of a given language, *SIAM J. Control Optim.*, **25**(3): 637–659.

Wonham, W. and Ramadge, P. (1988). Modular supervisory control of discrete event systems, *Mathematics of Control, Signals and Systems*, **1**: 13–30.

Yu, Y. and Gouda, M. (1982). Deadlock detection for a class of communicating finite state machines, *IEEE Trans. Communications*, **30**(12): 2512–2516.

Zeigler, B. (1989). DEVS representation of dynamical systems: Event-based intelligent control, *Proc. of the IEEE*, **77**(1): 72–80.

Zhong, H. and Wonham, W. (1990). On the consistency of hierarchical supervision in discrete-event systems, *IEEE Trans. Autom. Control*, **35**(10): 1125–1134.

Zuberek, W. (1986). Inhibitor D-timed Petri nets and performance analysis of communication protocols, *INFOR Journal*, **24**(3): 231–249.

Progress in Systems and Control Theory

Series Editor

Christopher I. Byrnes
Department of Systems Science and Mathematics
Washington University
Campus P.O. 1040
One Brookings Drive
St. Louis, MO 63130-4899

Progress in Systems and Control Theory is designed for the publication of workshops and conference proceedings, sponsored by various research centers in all areas of systems and control theory, and lecture notes arising from ongoing research in theory and applications control.

We encourage preparation of manuscripts in such forms as LATEX or AMS TEX for delivery in camera-ready copy which leads to rapid publication, or in electronic form for interfacing with laser printers.

Proposals should be sent directly to the editor or to: Birkhäuser Boston, 675 Massachusetts Avenue, Cambridge, MA 02139, U.S.A.

Related title: